From COWS to CONCRETE

The Rise and Fall of Farming in Los Angeles

Overleaf: Designed to lure small farmers to the region, the cover of a Los Angeles Chamber of Commerce booklet highlights agriculture as a key industry in Los Angeles. The promotional piece includes numerous photographs and glowing crop-by-crop descriptions of the bounty that was grown in Los Angeles County, touting, "An ample and unfailing water supply." 1929.

After the Gold Rush, *rancheros* drove hundreds and even thousands of cattle from Los Angeles to San Francisco to capitalize on high prices for beef. Los Angeles was famed for its skilled cowboys, or *vaqueros*, who struggled to keep the herds together as they faced challenges including fog, stampedes, and cattle theft. 1877.

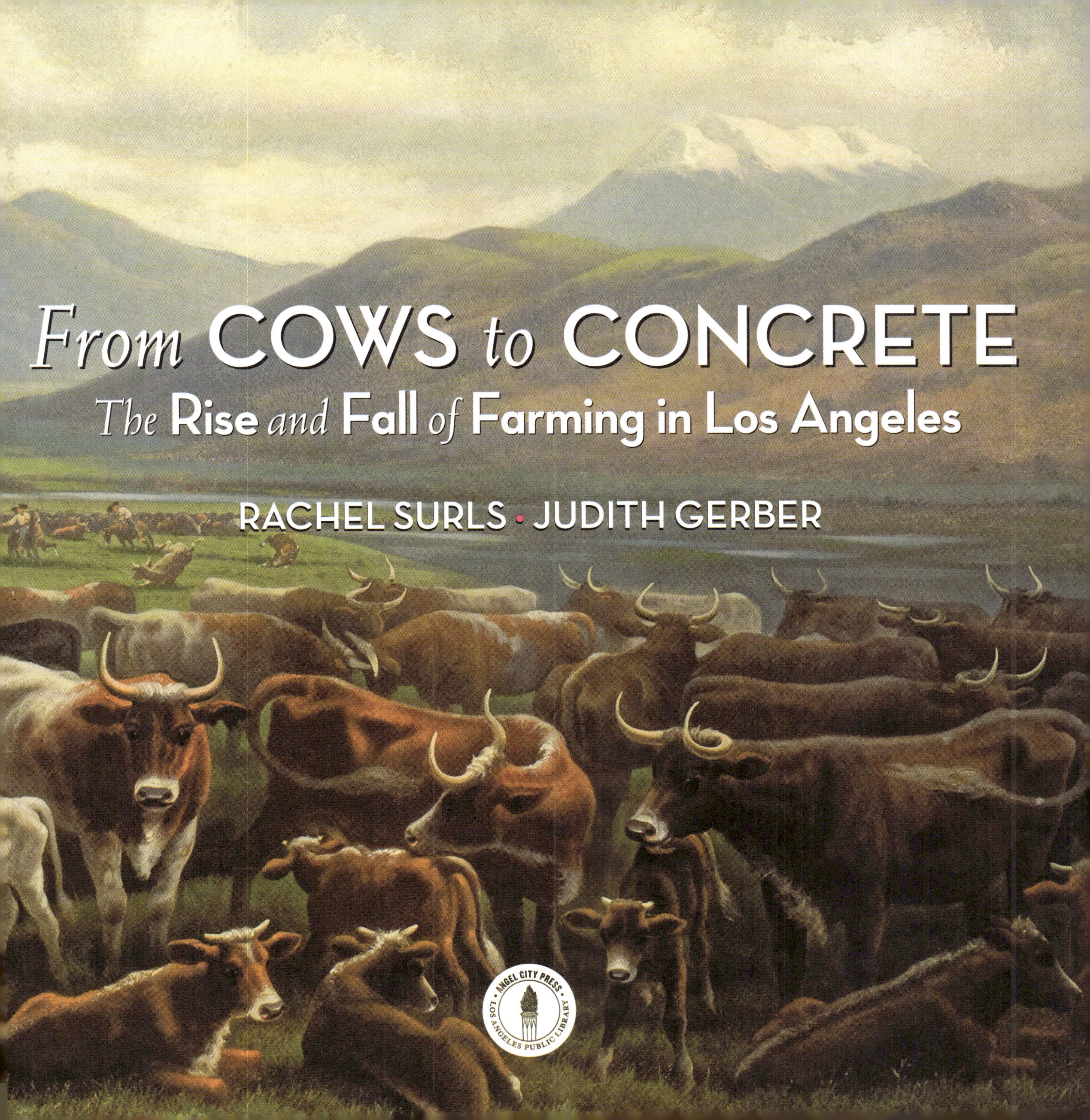

From COWS *to* CONCRETE

The Rise *and* Fall *of* Farming in Los Angeles

RACHEL SURLS • JUDITH GERBER

ANGEL CITY PRESS
LOS ANGELES PUBLIC LIBRARY

Ferdinand Deppe's oil painting of Mission San Gabriel Archángel depicts both the missionaries and Gabrieleño people (also known as the Tongva or Kizh), who became the mission's labor force. The California citrus and wine industries trace their roots to the mission. 1832.

From the native people of Los Angeles,
to the farmworkers who labored in its fields,
to the hopeful seekers who came in search of a new life,
this book is dedicated to those whose voices were lost along the way.

THE OLD
SPANISH AND MEXICAN
RANCHOS
OF LOS ANGELES COUNTY
MOJAVE DESERT
Los Alamos y Agua Caliente
La Liebre
Temescal
San Francisco
Simi
Tujunga
1797
Ex Mission de San Fernando
PUBLIC LAND
La Cañada
El Escorpion
El Encino
El Camino Real
Las Virgenes
El Conejo
Providencia
San Rafael
San Pascual
Santa Anita
Azusa de Duarte
Azusa Dalton
Addition to San Jose
Los Feliz
San Vicente y Santa Monica
San Francisquito
San Jose
San Jose de Buenos Ayres
Rodeo de Las Aguas
La Brea
Pueblo de Los Angeles
San Gabriel Arcangel
1771
Potrero Grande
Potrero de Felipe Lugo
Los Nogales
Topanga
Malibu
Sequit
Boca de Santa Monica
Rincon de Los Bueyes
Las Cienegas
La Merced
Potrero Chico
La Puente
La Ballona
Cienega o' Paso de la Tijera
San Antonio
Paso de Bartolo
Rincon de La Brea
La Habra
BAY OF SANTA MONICA
PACIFIC OCEAN
Aguaje de La Centinela
Sausal Redondo
Tajauta
Santa Gertrudes
Los Coyotes
San Pedro
Los Cerritos
Los Alamitos
Los Palos Verdes
Isla Santa Catalina
Prepared and Copyrighted by
Title Insurance and Trust Company
Los Angeles
Scale of English Miles
Spanish Leagues
Spanish Varas

Contents

◀ The Title Insurance and Trust Company created several *rancho* maps as a real estate marketing tool. The colored lithographs provided land-title abstracts as well as promoted title insurance. 1937.

Introduction

Near the intersection of Short and Poppy Avenues in Compton, the branches of a massive sycamore spread over an apartment complex in a neighborhood where few large trees grow. Known as the Eagle Tree, the old sycamore is a living remnant of Los Angeles County's largely forgotten history as a farming community. The Eagle Tree was one of the boundary markers of the giant Rancho San Pedro, a cattle ranch established in 1784 that sprawled over 75,000 acres from the Los Angeles River to the Pacific Ocean. Local ranches like San Pedro were self-sufficient, raising corn, wheat, and a few vegetables along with enormous herds of cattle. For the little they couldn't produce themselves, *rancheros* traded cowhides, building the first economy of Los Angeles.

Such pastoral history could not seem further away today. Compton has few gardens and little green space, which is true for much of Los Angeles County. Concrete covers the land where cows grazed and crops thrived. Under the pavement and parking lots of the vast urban landscape lie thousands of acres of once-productive farmland. Farming was at the center of life in Los Angeles from the time of its founding in 1781, an aspect of local history important well into the mid-twentieth century, when Los Angeles County was the top agricultural county in the nation.

Without farming, there would be no Los Angeles. The city exists only because in 1769 members of the first Spanish land expedition into Alta California recognized its potential for farming. Alta California, which encompassed the entirety of California today, was *terra incognita* for the Spanish, who claimed the

◄ Once a landmark for *rancheros*, the giant sycamore known as the Eagle Tree served to mark the northern boundary of the 75,000-acre Rancho San Pedro cattle ranch. 1952.

land, but had not explored beyond its coastline. In 1769, members of the group led by Gaspar de Portolá, traveling through the area as they made their way toward Monterey, observed that the native people were not farmers. Rather, they harvested the land's native plants. But the travelers noted the ingredients for successful farming: rich soil, lush plant life, and water nearby—plenty of water. In just over a decade, as part of Spain's effort to colonize this unknown land, Los Angeles was established as an agricultural village, on the banks of the river that would be its main source of water for more than a century.

Over the decades, Los Angeles grew from a small farming community into an agricultural powerhouse. Farmers experimented with a multitude of crops, from fruits and vegetables, to hemp, cotton, and flowers. Livestock was important too, with a major stockyard rivaling those in Chicago and Omaha, hundreds of dairies, and poultry ranches. Some enterprises faded away, while others thrived, influencing L.A.'s development as a metropolitan and cultural center. Local agriculture reached its apex in the four decades from 1909 to 1949, when Los Angeles County was the top farm county in the United States.

Farms in Los Angeles had influence well beyond the county's borders. From avocados to citrus to vineyards, several massive California agricultural industries first put down roots in Los Angeles County, helping to push the state into prominence as an epicenter for farming. Agriculture is still big business in California, the top farm state in the nation. And while it has faded from its glory days, Los Angeles County still makes a contribution to the Golden State's vast farm productivity. L.A. County farm production ranked thirty-second among fifty-eight California counties in 2013, with crops including alfalfa, carrots, and peaches, along with nursery crops. A catch-all category for shrubs, trees, flowers, and other plants produced for landscapes, nursery crops are one of the few types of commercial agriculture still relatively common in urban Los Angeles. Grown in containers, they can be squeezed into limited spaces such as land under power lines. The county's more traditional farms, on the other hand, are mostly in the sparsely populated High Desert, sixty miles north of downtown Los Angeles, where fields of bright green alfalfa and orchards of cherries and peaches still dot the countryside around the cities of Lancaster and Palmdale.

. . . if you know someone's ZIP code,

In the county's urban core, while farms are scarce, examples of the area's agrarian past abound, often in unexpected places. But more than just history, these examples are also relevant to the present and future. Often, they help to explain how Los Angeles evolved—sprawling and unwieldy, diverse and eclectic. The farm history of Southern California also suggests solutions for the future, as Angelenos engage with food and farming. Area residents have become passionate about their farmers markets, community gardens, and fresh foods in general. Local folks are fighting city hall on issues including the legality of backyard beekeeping and planting vegetables in the narrow strip of land between the sidewalk and the street. Urban farms are sprouting up, testing community tolerance for the sound of crowing roosters and the smell of compost. Some see these as meaningful connections with nature, others, as nuisances.

The current interest in farming is evident in other major metropolitan areas, too. But in L.A., these issues take on added importance because of its sheer size. What happens in Los Angeles has relevance for urban communities across the nation, in part because Los Angeles grapples with common issues on a much larger scale. Just as Los Angeles County was once a farming behemoth, it is the demographic heavyweight among counties in the United States, with more than ten million residents. For comparison's sake, the next largest county is Cook County, Illinois, which includes Chicago and its population of more than five million. With more than a quarter of California's residents, Los Angeles County has more people than all but eight states.

Los Angeles County is home to almost thirty percent of Californians who live in poverty. The level of income inequality has crept up over the past decade, according to the U.S. Census Bureau. There is enormous economic disparity in the City of Angels in particular, where the gap between rich and poor ranked ninth in the nation when the Brookings Institution reported on income inequality in 2014. The Los Angeles County Department of Public Health (LADPH) reported that the number of food-insecure households (meaning those consistently unable to afford enough food), had reached crisis proportions. Many families ran out of food before the end of the month, and hunger was all too common. According to a 2013 study conducted by a national coalition of food banks, Los Angeles County had more children at risk of going hungry than any other county in the entire United States.

you know their health.

Los Angeles, once famed as a land of health and sunshine, became known as a health-obsessed metropolis, with juice bars, vegan restaurants, and yoga studios galore—but not in every neighborhood.

"There's a saying if you know someone's ZIP code, you know their health," as one public health expert put it. East Compton, for example, had the highest adult obesity rate in the county, almost forty percent, while the affluent community of San Marino had the lowest, at just over eight percent, according to LADPH statistics released in 2011. In urban Los Angeles communities, burgers, fries, and unhealthy—but affordable—fast food prevailed in 2016. The land of fast food was home to an obesity epidemic and a resulting rise in chronic disease. Health experts and activists have connected the dots between public health and local farming, which some view as a strategy for making healthful food more available to urban residents. Hope springs eternal: in 2016, the celebrated chefs Roy Choi and Daniel Patterson teamed up to open a healthy and inexpensive fast-food establishment called Locol in Watts, which would bring fresh, local food to urban Los Angeles. Their goal was to take the concept nationwide.

Whether the reasons relate to health, the environment, or simply a desire to better understand food and how it's produced, farming has become part of the local and national dialogue. Once, Los Angeles was home to an extraordinary harvest, its bounty sustained for two-and-a-half centuries. What better time and place to start a conversation about the importance of honoring farm heritage? In that heritage is the story of those who cultivated the land the world calls Los Angeles—from the Native Americans who first gathered its wild crops, to a succession of immigrant newcomers who changed the cultural landscape.

From Cows to Concrete is the story of how thousands of acres of prime farmland disappeared under the spreading suburbs of the growing city. It's a cautionary tale, but also a story of hope, as urban residents in Los Angeles engage with the region's farm heritage. By connecting with the past, Angelenos hope to create a more abundant future, a Southern California where farms and gardens will once again take root amidst the concrete.

—R.S. and J.G., 2016

Without farming, there would be no

➤ The old sycamore, a majestic symbol of *rancho* days, was a lone reminder of the agricultural land of old. The tree is now gone, but a historic plaque marks its former location. Compton, 2011.

Los Angeles.

San Fernando Valley
Kokamohvit, Piwipwi
San Bernardino Mts.
Summit
Cajon Junction
Amutskupinga
Pasheeknga
Muhunga
Tuhunga
Sylmar
Porter Ranch
San Fernando
Sunland
Tujunga
Haramokngna
Chatsworth
'Ashaawanga
Achooykomenga
La Cresenta
Mount Wilson
Mount Blady
Yoat
Devore Height
Northridge
La Canada Flintridge
Aleupkinga
CHUMASH
Totonga
West Hills
Tarzana
Altadena
Toome Ahtsooingah
Santa Monica Mts
Burbank
Haramokngna
Akuuranga
Sierra Madre
Asuksanga
San Antonio Heights
Kuukaamonga
Cucamonga
Calabasas
Encino
Xaxaamonga
Pasadena
Glendora
Agoura Hills
Calabasas
Siutcanga
Sherman Oaks
Glendale
Sisitcanonga
Arcadia
San Bernardino
Valley
Cornell
Glassell Park
South Pasadena
Azusa
San Dimas
La Verne
Upland
Wa'atsnga
Fontana
Maaw'nga
Houtnga
Weniinga
Alhambra
Monte Nido
Beverly Hills
Los Feliz
Otsunga
El Monte
Covina
Claremont
Ontario
Bloomington
Komiicranga
Fernwood
Pacific Palisades
Century City
Yaanga
Los Angeles
Torojoatnga
Topa'nga
Kuruvanga
Apachianga
South San Gabriel
West Covina
Horuuvnga
Santa Monica
Boyle Heights
Isantcanga
Awiinga
Toibinga
Wajijanga
Malibu
Mar Vista
La Puente
Pomona
Glen Avon
Maalewonga
East Los Angeles
San Gabriel Valley
Chino
Mira Loma
Chokishnga
Pemookanga
Passinonga
Marina Del Rey
Huntington Park
Sejatnga
Whittier
Walnut
Los Serranos
Pedley
Pwinukipar
Inglewood
Maywood
Florence
Diamond Bar
Riverside
Playa Del Rey
Sa'anga
Bell Gardens
South Whittier
Nakaunga
Lennox
Downey
Norco
Paxauxa
Santa Monica Bay
El Segundo
Hawthorne
Willowbrook
Naxaawnga
La Habra
La Sierra
Casa Blanca
Norwalk
Los Angeles Santa Ana Plain
Compton
Chino Hills
Manhattan Beach
Huutnga
Amupunga
Paramount
Fullerton
Yorba Linda
Home Gardens
CAHUILLA
Lakewood
Cerritos
'Ongoova'nga
Redondo
Atwood
El Cerrito
Carson
Tibahanga
Anaheim
Hotuuknga
Torrance
Bixby Knolls
Cypress
Villa Park
Pamajam
Maasavnga
Palos Verdes Estates
Amaunga
Lomita
Orange
Cowan Heights
Shwaanga
Povu'nga
Rolling Hills Estates
Garden Grove
Malaga Cove
Long Beach
Westminster
Pasbenga
Rancho Palos Verdes
Seal Beach
Kingkingqaranga?
Silverado
Mutuucheynga
Santa Ana
Pt. Vicente
Tustin
Portuguese Bend
Moniika'nga
Alverhill
Toveemonga
Fountain Valley
White Pt.
Santa Ana Mts.
Pt. Fermin
Xoyuunga
Sunset Beach
Chaawvenga
Kiingkenga
Huntington Beach
Irvine
San Pedro
Chowiinga
Ataavyanga
Lopuuknga
Xaraashnga
Costa Mesa
El Toro
Newport Beach
Coto de Caza
kenyaanga
Mission Viejo
Moyonga
Corona Del Mar
ACJACHEMEM
Laguna Beach
JUANEÑO
Emerald Bay
San Pedro Channel
'Tchunashnga
San Juan Capistrano
Santa Barbara Is.
Luiseñ
Naayxoxar
(Isthmus)
Dana Point
Capistrano beach
Pimu'nga
San Clemente
Santa Catalina Is.
Avalon
Xaraashnga
San Nicolas Is.
The Four Sacred Mts. Wahchah Ahhoov Hihieesh
Mt. Baldy Joat or Yoat
Mt. Saddleback Har'wovet
Mt. San Gorgonio Akvagna
Mt. San Jacinto Jamiwu
The Four Sacred Rivers Wahcha Ahhoov Pahpahhite
Los Angeles River Páymi pahit(e)
Hondo River Chenóoeh pahhít(e)
San Gabriel River Totótingah Pahhit(e)
Santa Ana River Kahho Pahhít(e)
N
Kiingkenga
San Clemente Is.
Kiinke Paar
Villages
Sacred Sites or Villages?
Villages in ???
Villages & archceological sites

1 Before Los Angeles

Spreading over thirty-two acres at the edge of Chinatown, the Los Angeles State Historic Park is a rare stretch of open land in the heart of the city. The site provides a striking example of how land has transformed over four centuries, as well as a vantage point to explore what Los Angeles was like before it was the heart of a metropolis. It's near the site of Yaanga, the native American village whose residents greeted the first Europeans to visit the area. Settlers in the early Pueblo of Los Angeles planted grain, orchards, and vineyards in its fertile soil. As the pueblo grew into a city, the land was home to the first station for the Southern Pacific railroad that connected Los Angeles with the rest of the country. According to local lore, corn seeds fell off railcars and sprouted along the tracks, resulting in an enduring nickname for the land that's now a park: "The Cornfield."

While much of the Cornfield's past is invisible today, it does include a tangible remnant of early Los Angeles: a section of its first water system, the Zanja Madre, or mother ditch. The *zanja* was the lifeline for the farming village first known as El Pueblo de la Reyna de Los Angeles, bringing water from the Los Angeles River, then called El Río de Nuestra Señora La Reyna de Los Ángeles de Porciúncula.

Only because of the river did Spanish explorers stop to cool off and set up camp on a summer day in 1769. Father Juan Crespí, a member of the group, wrote that it was a "delightful place, among the trees on the river." The abundance of the land, with the river in its midst, was the main reason

Jacob Gutierrez created this twenty-first-century map of pre-colonial villages in the Los Angeles basin. He is a member of Keepers of Indigenous Ways, a group of descendents of residents of the Shwaanga village in the Harbor City/Wilmington/Lomita area. Over the years, the native people of the Los Angeles area have been known as the Gabrielino or Gabrieleño, and more recently as the Tongva. However, some historians have questioned the authenticity of the name "Tongva" and suggest that "Kizh" is a more historically accurate name for the native people of Los Angeles. c. 2008.

▲ A solitary Gabrieleño woman sits in front of a tule-and-willow shelter typical of those built in the years after colonization. c. 1890.

the Spanish came to settle the place that would become Los Angeles. On this first journey through the area, they immediately saw its potential as valuable farmland.

"After crossing the river we entered a large vineyard of wild grapes and an infinity of rosebushes in full bloom," wrote Crespí. "All the soil is black and loamy, and is capable of producing every kind of grain and fruit which may be planted. . .. This plain where the river runs is very extensive. It has good land for planting all kinds of grains and seeds, and is the most suitable site of all that we have seen for a mission, for it has all the requisites for a large settlement." His observation was prophetic. In just over a century, the area around the river became the center of a farming empire. Later, it would be at the core of an enormous, sprawling metropolis.

Crespí and the other travelers were keenly interested in productive land, since farming was key to Spain's plans for Alta (upper) California, the unexplored territory above Baja (lower) California. Gaspar de Portolá led the advance team for colonization whose mission was to find a land route to Monterey, which was to become the first military outpost in this unknown land.

Portolá's expedition came after almost three centuries of Spanish colonization of the New World, driven by a quest for gold and other riches. Beginning in 1496 with the first settlement on Hispaniola, the island now home to Haiti and the Dominican Republic, the Spanish conquest spread across the Caribbean, then into the Mexican mainland in 1519. A royal court was established in Mexico City in 1527. There, the colonial government of New Spain took root, as the conquest spread north through Mexico and south through Central and South America, toppling ancient empires.

By the 1530s, the Spanish colonial government had California in its sights. Explorers believed that somewhere along its coastline lay the hidden entrance to a waterway that crossed the North American continent. This route, if discovered, would provide a shortcut from Spain to Asia. They borrowed the name for this unknown land, California, from a popular Spanish novel about the exploration and conquest of a mysterious island, *The Labors of the Very Brave Knight Esplandián*, written in 1510 by Garci Rodríguez de Montalvo.

"I will tell you that on the right-hand side of the Indies there was an island called California, which was very close to the region of the Earthly Paradise," wrote Rodríguez de Montalvo. "This island was inhabited by black women, and there were no males among them at all, for their way of life was similar to that of the Amazons. The island was made up of the wildest cliffs and the sharpest precipices found anywhere in the world." In this imagined place, the warrior women wore golden armor. Griffins—the half-eagle, half-lion beasts of Greek mythology—defended them, devouring any man who dared attack. The swathe of land called California, after this fabled island, was, like its namesake, beautiful, mysterious, and potentially lethal. And in fact, the Spanish believed that California was an island until exploration proved otherwise.

An early-twentieth-century Gabrieleño seed beater made from willow was found in Pacific Palisades. Seed beaters were used to harvest seeds from plants while leaving the plant intact, taking only seeds that were needed for food and leaving some to fall and reseed. c. 1900.

Infamous conquistador Hernán Cortés explored the coastline of Baja California by sea in 1535, although it took a century and a half before a permanent settlement was established at Loreto. In 1542, Juan Rodríguez Cabrillo sailed north from Mexico to explore the coastline of Alta California. Though Cabrillo claimed the land for Spain on this voyage, his party never ventured far from shore. The interior remained unknown to Europeans for more than

A portion of the Zanja Madre was unearthed at the Los Angeles State Historic Park in 2000. The *zanja* brought water from the Los Angeles River, then simply referred to as the Porciúncula, for drinking, bathing, and irrigation. 2000.

two centuries. Meanwhile, as officials in Mexico City postponed exploration into Alta California, eighteen missions were established up and down the eight hundred-mile length of Baja.

Sightings of Russian vessels along the Alta California coast finally prompted Spanish officials to secure their claim to the territory. Led by José de Gálvez, the visitor-general of New Spain, they planned to follow the standard colonization strategy. They would set up military bases, or presidios, to protect their territory, while building missions to convert the natives to Christianity, creating a local workforce. Nearby civilian pueblos were planned, which would grow food to supply the needs of the colonists.

Gálvez decided that the initial expedition would leave from Baja California and establish a base of operation at San Diego (named in 1602 by explorer Sebastián Vizcaíno). From there, the expedition would move north to Monterey to create a presidio. Gálvez appointed Portolá, an experienced soldier and the current governor of Baja, to command the expedition. He named Father Junípero Serra, the new leader of the Baja missions, to take charge of mission development in Alta California. The expedition would include two parties traveling by land, and two by sea.

Portolá and Serra, traveling with the second land contingent, left Baja on May 15, 1769 and arrived in San Diego on July 1. Within two weeks, Portolá decided to press on to find a land route to Monterey, while Serra stayed behind to establish a mission at San Diego (Serra later became famous as the "father" of the California missions and was beatified in 2015 by Pope Francis, who was criticized by many for this decision because of Serra's treatment of Native Americans). Crespí, Serra's former student and the designated recordkeeper for the expedition, left with Portolá to document the trip north.

As they traveled, Crespí and the other members of Portolá's party took note of the land and water, since productive farms were key to the three-pronged colonization strategy of establishing missions, presidios, and pueblos. They also interacted with the native people who lived along their path, whom they viewed as the necessary labor force for settling the region.

According to Crespí's journal, the native people were friendly, curious, and generous. When Portolá's company set up camp by the river on the second of August, in the place that would become Los Angeles, the residents of the nearby village paid a visit. "As soon as we arrived, about eight heathen from a good village came . . . They presented us with some baskets of *pinole* made from seeds of sage and other grasses. Their chief brought some strings of beads made of shells, and they threw us three handfuls of them. Some of the old men were smoking pipes well made of baked clay, and they puffed at us three mouthfuls of smoke. We gave them a little tobacco and glass beads, and they went away well pleased."

The "heathens" who greeted the Portolá party were the predominant Native American tribe throughout much of the land now known as Los Angeles County and northern Orange County, a territory that stretched over more than 1,500 square miles, and included the islands of Catalina and San Clemente. Although it is unclear what they called themselves before their contact with Europeans, today, these people are known as the Gabrielino or Gabrieleño, named for the San Gabriel Mission, or sometimes, as the Tongva or Kizh. Approximately five thousand Gabrieleño lived in fifty to a hundred small communities, like Yaanga, the village whose residents greeted Portolá's party and presented them with baskets of seeds.

The seeds they offered to the travelers represented sustenance and survival. Like other Gabrieleños and native people of Alta California, the residents of Yaanga were not farmers, but gathered the land's wild harvest. The seeds of local grasses and other plants, painstakingly collected with special woven beaters and baskets, were an important food source. This place that would become Los Angeles was a prairie, lush with grasses, clovers, and wildflowers, many of which produced edible seeds, for example, chía, used to make flour and beverages. Roots, bulbs, and tubers of various plants, such as wild hyacinth, nutsedge, Humboldt Lily, and wild onion were other key harvests, dug up with long, pointed sticks, then baked or roasted.

In the foothills, the land transitioned to chaparral and forest, full of shrubs and trees abundant with food, including oak trees. Acorns were the most important crop for the Gabrieleño and many other Alta California people, despite being naturally bitter and barely edible without special handling. However, native people had discovered that the bitterness could be removed by processing acorns in various ways, such as grinding them, then leaching the meal repeatedly with hot and cold water. Acorn meal was then made into a mush that provided a nutritious dietary staple. And unlike most staple crops today, acorns did not need to be planted every year, but rather, were harvested from trees that lived for centuries.

Gabrieleño took advantage of plentiful game, as well as fish and shellfish along the coast. Reporting on native cultural practices for the *Los Angeles Star* in 1852, Scotsman Hugo Reid, who had married a Gabrieleño woman, wrote: "The animal food in use among them was deer meat, young coyotes, squirrels, badgers, rats, gophers, snakes, raccoons, skunks, wildcats, the small crow, the blackbirds, hawks, ground owls [sic], and snakes, with the exception of the rattle snake [sic]. A few eat the bear, but in general it is rejected, on superstitious grounds. The large locust or grasshopper was a favorite morsel, roasted on a stick at the fire. Fish, whales, seals, sea-otters, and shellfish, formed the principal subsistence of the immediate coast-range of Lodges and Islands."

Although the food supply was often ample, there were also lean times. An important role of the Gabrieleño tribal chief, known as the *tomyaar*, was to manage and protect community food resources. In the Gabrieleño culture, which discouraged gluttony and promoted sharing, all were expected to contribute a portion of their harvest or hunt toward a communal pool.

The native people of Alta California, while not farmers, sometimes encouraged the growth of their wild crops. It's likely that the Gabrieleño did so as well. For example, some California tribes collected seeds of their preferred wild plants, then scattered them in spots where they wanted them to grow. They pinched off new shoots of plants, ensuring branching and new, healthy growth. Firsthand accounts of Crespí and others document that the Gabrieleño (like other Alta California tribes) set fires to manage the land, using carefully controlled burns to rid their gathering areas of pests and promote new growth.

Gathering, hunting, and encouraging a bountiful wild harvest were based on the seasons and the cycles of nature. However, the Spanish invaders viewed this lack of a daily, regimented schedule as a sign of laziness. They would soon build missions to "civilize" the Gabrieleño, forcing them to join a workforce that would further the goals of the church and the colonial government.

Through the process of colonization, the world of the Gabrieleños would be permanently altered. Thousands of native people would die, many falling victim to diseases introduced by the colonists, while others would leave the area, fleeing the control of local missions. Those who remained would become integral to the development of the tiny pueblo that became Los Angeles. A small handful would become landowners, but most would be the first farm laborers at the missions, in the Pueblo of Los Angeles, and on the surrounding *ranchos*. The land would change as well, impacted by the colonizers and their herds of grazing cattle that destroyed the native vegetation. The Gabrieleño's lightly managed landscape, with vast prairies of wildflowers and native grasses, a warm climate, rich soil, and access to water, was viewed by the newcomers as a commodity to be exploited. In just a few decades, colonization would consume and transform the land forever.

Yaanga, the village that existed before Los Angeles, vanished into history. Its location is unclear, but may have been in the vicinity of today's City Hall. A mile away, the Cornfield, site of the Los Angeles State Historic Park, survived obliteration. After a century as a railyard, the land sat vacant until an influential developer chose it as a site for a complex of warehouses. A neighborhood coalition rose up to oppose the project, resulting in a rare win for the community, which would get a park instead of another development. Yet the land, once so fertile, was now dusty and desolate, full of shopping carts, broken televisions, and other urban detritus.

In 2005, a local artist, Lauren Bon, transformed the site into a temporary art installation that highlighted the land's former abundance and played on its longtime nickname. After removing tons of trash, Bon's Metabolic Studio trucked in thirty thousand cubic yards of topsoil and sowed more than eight hundred thousand corn seeds to create the exhibit. Titled *Not a Cornfield*, it was, in fact, an enormous cornfield, representing the first time the land had been used to grow crops since 1876, when it became the site for the Southern Pacific train depot. Intrigued, Angelenos watched the corn, as it sprouted, grew tall and green, then turned golden as harvest approached. Although it existed for just one season, *Not a Cornfield* served as a sly reminder that Los Angeles was once a place of bountiful crops. And perhaps, it could be, once again.

The 2005 *Not a Cornfield* temporary art installation at the Los Angeles State Historic Park was planted as a visible reminder of the agricultural heritage of the thirty-two-acre site. City Hall can be seen through the stalks. 2005.

LEGALIZE
HEMP
#1 SOURCE OF
MEDICINE
PLASTIC
FOOD
hemp
LEGALIZE
HEMP
#1 SOURCE OF
PAPER
FIBER
FUEL
MEDICINE
PLASTIC
FOOD
SIGN UP HERE
DONATIONS TO HELP LEGALIZE HEMP
HEMP
NON-TOXIC
BIODEGRADABLE
RECYCLABLE
RENEWABLE
FUEL · FABRIC · FOOD · MEDICINE · PLASTICS

2 Farming in Colonial Los Angeles

Los Angeles has a long history with *Cannabis*, the plant primarily known as marijuana these days. It's readily accessible, thanks to medical marijuana clinics and delivery services. It's highly visible, with events like "Kush Expo" and "HempCon" drawing crowds at the giant L.A. Convention Center. But whatever you call the product—weed, pot, chronic, or kush—it's a crazy quilt of legality.

California law allows marijuana use for medical purposes, yet federal law prohibits it, no matter the circumstances. It's no wonder that the Los Angeles City Council struggles with regulation of the marijuana dispensaries found throughout the city, each marked with the distinctive green cross. Three separate initiatives related to storefront pot sales were on the ballot in the city's 2013 elections, and voters chose City Council-sponsored Measure D, which limited the number of dispensaries. Before and after, some wondered aloud whether politicians and advocates had over-consumed the substance in question, and could have better addressed this matter in the state legislature.

Cannabis was also a hot political issue in the earliest days of the Pueblo of Los Angeles. At that time, the concerns were about hemp, the other *Cannabis* breed. (Hemp and marijuana are the same species, but marijuana varieties are bred to have more tetrahydrocannabinol, or THC, the chemical responsible for the mind-altering high; hemp has traditionally been grown for its fiber, much like cotton.) The first cash crop in Los Angeles, hemp was promoted by Spanish authorities who hoped that a profitable crop would shore up their efforts to colonize the Los Angeles area.

◄ A stand on the Venice Boardwalk encouraged voters to support California's 1994 initiative to legalize hemp.

In 1804, at the time hemp was introduced, the Pueblo of Los Angeles had existed for just over two decades, but had only 315 residents. Few wanted to come to this remote outpost, so far from any city. Nearby, the Mission San Gabriel Archángel, established in 1771, struggled with a rebellious native population. Thirty-five years after Portolá's expedition, whether colonization would succeed was an open question.

The Spanish colonial government hoped that hemp would be a solution. Valued for making high-quality rope used for ship riggings, hemp was an important product for a naval power like Spain. To promote hemp farming, Spanish officials offered an inflated price as an incentive. They even sent an expert to help the colonists grow the crop—Joaquin Sanchez, a naval gunner who had grown hemp in Spain. The experiment succeeded. By 1808, Los Angeles produced seventy percent of the Alta California hemp crop. Not all pueblo citizens agreed with its production, however. Some petitioned the governor, complaining that thirsty hemp plants were soaking up water needed for their wheat and corn. The governor declined to intervene since he considered hemp production key to Alta California's success.

At that point, Los Angeles—or as the Spanish colonists called it, El Pueblo de la Reyna de Los Angeles—was still in its infancy, struggling to fulfill its intended role as a farming center. The second civilian pueblo in Alta California, it was established in 1781, just four years after the Pueblo of San José. Like San José, it was founded as a civilian agricultural village, tasked with supplying the presidios, or military forts, that protected the fledgling Alta California colony. In fact, the new governor of the Californias, Felipe de Neve, had developed the *Reglamento,* detailed guidelines on exactly how pueblos such as Los Angeles would be established, including rules for farming.

"To each settler . . . must be given two mares, two cows with one calf, two ewes, and two she-goats, all pregnant, and one yoke of oxen or bullocks, one ploughshare or tip, one hoe, one spade, one ax, and one sickle, one field-knife, one lance, one musket, and one dagger, two horses and one cargo mule," Neve wrote in the *Reglamento*. He dictated that residents would share outlying lands for pasturing their animals and harvesting wood, contribute grain to a communal granary, and work together to construct and maintain the pueblo's irrigation ditches, dams, and reservoirs. Each settler would receive four fields, two for irrigated farming, and two for dry farming. Using today's measurements, each of these individual fields would be a square 550 feet on a side—or the length of about one and a half football fields, squared.

Neve planned the new Pueblo of Los Angeles for a site along the Porciúncula River, near the spot where Father Crespí had admired the fertile land along the river's banks in 1769. To make his planned pueblo a reality in this remote place, Neve needed settlers, so he sent his Lieutenant Governor Fernando Rivero y Moncado on an expedition to recruit settlers and soldiers from the Mexican mainland in Sinaloa and Sonora, more than a thousand miles away. His recruits ultimately included eleven families, forty-four people in all. The men selected were those who would bring important skills to the new community—farmers, blacksmiths, carpenters, and masons—and could be enticed with promises of land, no matter how far away. Hundreds of animals came with them, including mules, oxen, cattle, and sheep, herded along the thousand mile trek to stock farms and ranches in the new settlement. After a difficult journey that included a smallpox outbreak and an attack by native people that killed several members of the party including Rivero y Moncado, the settlers made their way to the site of the new pueblo. September 4, 1781 was noted as the date of its official founding. By October, Neve's settlers had constructed an irrigation ditch, or *zanja*, from the river to their fields, where they were already planting crops.

A year later, however, Neve was less than pleased with the progress. He had been promoted to a new position, commander general of the internal provinces, and left written instructions for his replacement to keep an eye on the settlers. "It is most essential that special watchfulness be maintained with respect to the pueblo of Nuestra Señora de Los Angeles. Its location, like that of the town of San José, its plentiful water, land, and everything else necessary for progress, are advantageous, but it is necessary to have an active and exacting man there who will bestir the settlers to cultivate their lands, care for their crops, and do everything

▼ A view of the south façade of Mission San Gabriel with the mission's adobe house on the left. The adobe was not only home to the mission priests, but housed the granary, carpentry shops, and weaving rooms. c. 1880.

else connected with farming. The slight care they took in watching over the wheat harvest reduced the yield from a possible four hundred *fanegas* to 260. Through the same neglect, they lost their first crop of corn. The plants, already sprouted, dried up because of their carelessness in not opening up the irrigation ditch soon enough to water them," complained Neve upon his departure. (The term *fanegas* was a measurement of yield similar to a bushel. The yield decline he described would have represented a drop from about 640 bushels down to 416 bushels.)

As Neve realized, water was the key to successful farming in Los Angeles, the fine edge between a successful harvest, and a crop that shriveled and died. The settlers continued to develop the open irrigation ditch known as the Zanja Madre, diverting it to smaller ditches reaching individual fields. Gravity fed, the Zanja Madre started from a high point above the pueblo. A *toma* (dam) was constructed to divert river water into the ditch. Later, a waterwheel and more elaborate pipe system was added.

Whether digging irrigation ditches or planting grain, the settlers looked to the local village of Yaanga for labor, typically paying native workers in goods such as cloth, corn, and tools. The relationship between the Gabrieleños and the pueblo residents was relatively peaceful, especially when compared to the ongoing conflicts between the native people and the

Part of Mission San Gabriel's original vineyard still survives in the back of the mission. Date unknown.

local mission personnel. After Portolá's exploratory journey through the region in 1769, it was just two years before a group of two priests, ten soldiers, and a handful of settlers arrived to start a mission, the first Spanish settlement in the Los Angeles area. While Portolá's party found the native people friendly and welcoming, the group arriving in 1771 received a far less favorable reception, encountering Gabrieleños "in full war-paint and brandishing their bows and arrows, with hostile gestures and blood-curdling yells," according to Father Benito Cambón, one of the priests in the group. This early conflict, though quickly resolved, foreshadowed the volatile relationship between the missionaries and the Gabrieleños over the next four decades.

The Mission San Gabriel Archángel, founded on September 8, 1771, was the fourth Alta California mission. Conflict arose within weeks of the mission's establishment after a soldier raped the wife of a local *tomyaar*, who then led an attack on the mission, only to be killed, his head displayed on a stake inside the mission stockade. This incident led to a series of attacks by the Gabrieleños. It was months before more than a handful of native people were again willing to interact with the mission personnel.

Over time, though, many Gabrieleños came to the mission willingly to become converts, known as neophytes. Curiosity, a secure source of food, as well as offers of clothing and other items helped to attract neophytes. Once they had undergone conversion, however, there was no leaving without great risk, since neophytes who attempted to return to their villages or flee the area were often pursued and captured by mission soldiers, then locked in stockades or lashed. Periodically, the native people rebelled, most famously in 1785, when Toypurina, a twenty-four-year-old female shaman, led a failed attempt by a group of her people to regain control of their land and lives.

Despite a potentially mutinous workforce, as well as challenges with unfamiliar weather and pests, farming at San Gabriel was successful; its fields were soon out-producing other Alta California missions. Mission San Gabriel Archángel, which has been called the "Mother of California Agriculture," was the testing ground and incubator for what would one day become massive agricultural industries. Mission orchards, kitchen gardens, and grain fields served as experimental plots to learn how crops fared in the new land.

All of this was accomplished with Gabrieleño labor. Their native food sources had thrived with limited intervention, but the colonists introduced crops that required a completely different approach. Wheat, corn, and barley were the key crops for colonization, since they could easily be stored and traded. For these, land had to be prepared, using tools such as simple plows, hoes, and pick-axes to turn over soil that had previously been disturbed only by digging sticks and periodic burning. Land that was lightly managed for centuries was now transformed from prairie to farmland.

Another change was the introduction of irrigation, which was not part of native culture, but had been used in Spanish civilization for centuries. The mission personnel employed this knowledge at Mission San Gabriel, engineering canals and ditches to move water from rivers and streams into fields. Crops such as apples, grapes, oranges, figs, pears, and a wide variety of vegetables and herbs were cultivated in the mission's flourishing gardens, made abundant through irrigation.

While fields of wheat, corn, and barley were the main focus of mission agriculture in the early years, over time, ranching became the mission's most important business. Variable weather, with flooding some years, and drought in others, made grain harvests unreliable. Ranching was more consistent. Herds of cattle and flocks of sheep multiplied rapidly on the grassy plains of the Los Angeles basin, producing cowhides and wool that could be sold or traded, along with meat for the mission tables. At the peak of the mission's ranching endeavors in 1819, there were 16,000 head of cattle and 13,000 sheep on the mission's ranches that sprawled over 1.5 million acres, stretching from the coast to San Bernardino.

But ranching devastated native hunting and harvesting areas, as cattle and sheep overgrazed the land, causing erosion, changing water tables, and transforming the ecology of the Los Angeles basin. What had been a vast prairie, filled with diverse native plants, was drastically altered. The abundant rabbits, ground squirrels, and even grasshoppers, all important food sources for the original residents of Los Angeles, were soon viewed as agricultural pests. Grazing not only destroyed native plant life that had sustained the Gabrieleños for generations, it also restricted access to traditional trade routes and water sources, leaving many native people with little choice but to turn to the mission.

And while it promised security, life at the mission meant confinement in close quarters, with poor sanitation, and a limited diet of starchy food. Under these conditions, many succumbed to measles, tuberculosis, influenza, syphilis, and other diseases they had not been exposed to. The average death rate at Mission San Gabriel was ninety-five deaths per thousand people per year, while the birth rate was only forty-four per thousand. The statistics were grim throughout the Alta California missions, with devastating losses to the population of native people living at missions.

Mission San Fernando Rey de España's Convento Building (or long building) served as a hotel for travelers and visitors. At its peak, Mission San Fernando had 12,000 head of cattle and 7,800 sheep, along with a seventy-acre olive grove that served as the foundation of the San Fernando Valley's future olive industry. The building was completely restored, the only original building left at the mission. c. 1870.

Despite the high mortality rates among neophytes, the Los Angeles area continued to missionize with the founding of Mission San Fernando Rey de España in 1797, twenty-two miles northwest of the pueblo in the San Fernando Valley. The new mission proved to be a productive farming site as well, with an abundant harvest of corn, wheat, beans, garbanzos, lentils, and peas, as well as a seventy-acre olive grove that planted the seeds of a future olive industry in the San Fernando Valley.

Two crops thrived at the missions that would later form the basis of massive industries. Wine grapes were important at both Los Angeles-area missions, with grapes first planted at San Gabriel in the late 1780s. The mission padres used wine for mass as well as at their tables, where they entertained visitors. The success of wine grape production at the two missions would one day serve as the foundation of a huge and successful Los Angeles wine industry, the first in California. The California citrus industry had its roots at the Mission San Gabriel, where the first significant grove of orange trees in Alta California was planted around 1804 or 1805.

By 1810, the fragile colonization effort in the Los Angeles area was solidifying, thanks in part to productive farming in the pueblo and at the missions. Yet all this was at risk due to events beyond Alta California. In 1810, Mexico rebelled against Spain, sparking a decade-long struggle for independence. The war had major implications for Los Angeles, since ships no longer came from Mexico to deliver supplies. The tiny pueblo and local missions had to become self-sufficient in order to survive.

The stage was set for another era. Power had shifted away from the missions, in part because the local residents resented the amount of land held by the church. A new system of *ranchos* outside of the missions had already begun to take root. Spreading from the coast, to the mountains, to the inland desert, independent *ranchos* would become the predominant agricultural system in Los Angeles for years to come.

3 The Rise and Fall of the *Ranchos*

Bruno the Steer lives the good life, nibbling on a bale of hay in his spacious enclosure at the Farm Sanctuary, an animal rescue in northern Los Angeles County that provides a home for abandoned and abused livestock. Visitors touring the farm, including a flock of small children, scratch his ears and pet his enormous flank while Bruno sits patiently. A tour guide describes his temperament as similar to that of a Labrador Retriever, friendly and tolerant. Bruno, whose rescuers oppose the use of animals for food, is one of relatively few cattle in Los Angeles County. This might not seem surprising, except that Los Angeles was once known as the "Queen of the Cow Counties" because of its prolific cattle production. Until the 1860s, impressive cattle herds could be found grazing throughout Los Angeles County, presided over by the *rancheros* of the Spanish and Mexican eras. If Bruno had been around then, he would be running on the open range, lassoed and branded by Los Angeles's famed *vaqueros*, and eventually, slaughtered for his hide.

As the mission priests had learned, ranching was the most practical way to extract a profit from the grassy prairie of the Los Angeles basin.

While beef is the most important product from cattle today, in early Los Angeles, cowhides were the main reason to raise cattle, so fundamental to the local economy that they were known as "California bank notes." *Rancheros* traded their hides for cloth, furniture, sugar, whiskey, and other goods with the Yankee trading ships anchored off San Pedro. Transported to Boston factories, Los Angeles hides were

◄ Sheep and cattle graze on the banks of Johnston Lake in Highland Park. Even though ranching declined sharply with the end of the *rancho* era, some farmers still raised cattle and sheep. c. 1900.

manufactured into leather shoes, boots, and saddles.

Richard Henry Dana's classic 1840 memoir of life on a trading ship along the California coast, *Two Years Before the Mast*, describes the business of hides and their handling. After *rancho* workers slaughtered cattle, they stripped the flesh off the hides, dried them in the sun, then folded them in half lengthwise, with the hair inside. *Rancheros* sent their dried hides by oxcarts to the primitive port at San Pedro, twenty-five miles south of the Pueblo of Los Angeles. Dana wrote of San Pedro, "to my surprise. . .the desolate-looking place we were in was the best place on the whole coast for hides." The mild climate off the coast helped grass and other plants grow, which in turn nourished the largest herds of cattle in California and produced the best hides. After unloading the carts, Dana and other workers threw the bulky hides down the bluff, then carried them over their heads through the surf to the small boats that ferried cargo out to larger ships.

Cattle provided other critical products as well, such as tallow, rendered fat that was used to make candles and soap. Rawhide was also important, serving as the duct tape and bungee cord of the day, used to make quick repairs and easily tie things together. In today's beef-obsessed culture, where burgers and steaks dominate the culinary landscape, it's notable that beef was a much less important product of early cattle ranching than the hides and tallow. While it was certainly on the daily menu at *ranchos*, beef did not become important economically until the Gold Rush. Before that, it was mostly a byproduct of cattle ranching; stripped off the carcass, dipped in salt water, and dried in the sun, the resulting jerky was often cooked with water and potatoes to make stew.

Cattle cost little to produce, since the land was essentially free, assigned to settlers and retired soldiers in the hope that they would stay and cultivate a place that was so far from their known world. Assigning land to loyal subjects of the crown was a custom that dated back to Spain, a strategy often used during Spanish colonization of the New World. Governor Pedro Fages, who had marched with Portolá on his 1769 expedition to Monterey, made the first land grants in Alta California. In the 1780s, Fages assigned vast tracts in the vicinity of San Gabriel Mission to three soldiers who had served under his command. These first *ranchos* began a system of land distribution that would have a profound impact on the development of Los Angeles and the rest of California.

Fages granted the largest of these land concessions, comprising almost three hundred thousand acres, to retired soldier Manuel Nieto. Fages later cut this massive parcel nearly in half when the priests at the mission complained that it took away too much of their grazing land. Even so, the remaining land, known as Rancho Los Nietos, covered much of what would one day become several large cities in Los Angeles County: Long Beach, Downey, Lakewood, Norwalk, Santa Fe Springs, and Whittier, as well as a sizeable chunk of modern-day Orange County.

Manuel Domínguez, grandson of Juan José Domínguez, developed Rancho San Pedro and built the Domínguez *rancho* adobe as its headquarters. The Dominguez Adobe in Carson is California Historical Landmark No. 152 and was entered in the National Register of Historic Places in 1976. Date unknown.

The governor granted Juan José Domínguez, another soldier from his command, a 75,000-acre tract known as the Rancho San Pedro, which spread from the Los Angeles River at its eastern boundary, to the Pacific Ocean on the south and west. Its northern boundary continued inland toward what is now Compton. Fages made a third grant to José María Verdugo, a 36,000-acre piece of land called Rancho San Rafael. Though it was the smallest of the three allotments, parts of Rancho San Rafael would one day become Glendale and the northeast portions of the City of Los Angeles.

What started in Los Angeles spread through Alta California, with territorial officials assigning some thirty massive tracts of land during the Spanish colonial period, mostly clustered near Los Angeles, Santa Barbara, Monterey, and San Francisco. The number of *ranchos* slowly multiplied after many of the original grantees carved up their land for their heirs.

Life on early Los Angeles *ranchos* was gritty and unglamorous: families lived in tiny two-room adobe homes with thatched roofs, dirt floors, beds made out of leaves, and hides for blankets. Far from any real population center, *rancheros* had to be self-sufficient. "They planted and harvested the things they needed most for the benefit of their families, such as rice, corn, beans, barley and other grains, squash, watermelons and cantaloupes," according to José del Carmen Lugo, who grew up on a Los Angeles *rancho* in the early 1800s.

Over the years, though, through trading hides from their expanding herds, the *rancheros'* lives became easier, more luxurious, as they delegated work, often to Native Americans. The community of workers expanded over time, with Mexican settlers intermarrying with native

A Federal Writer's Project map of southern L.A. County juxtaposes 1937 towns and cities with the Mexican and Spanish land grants.

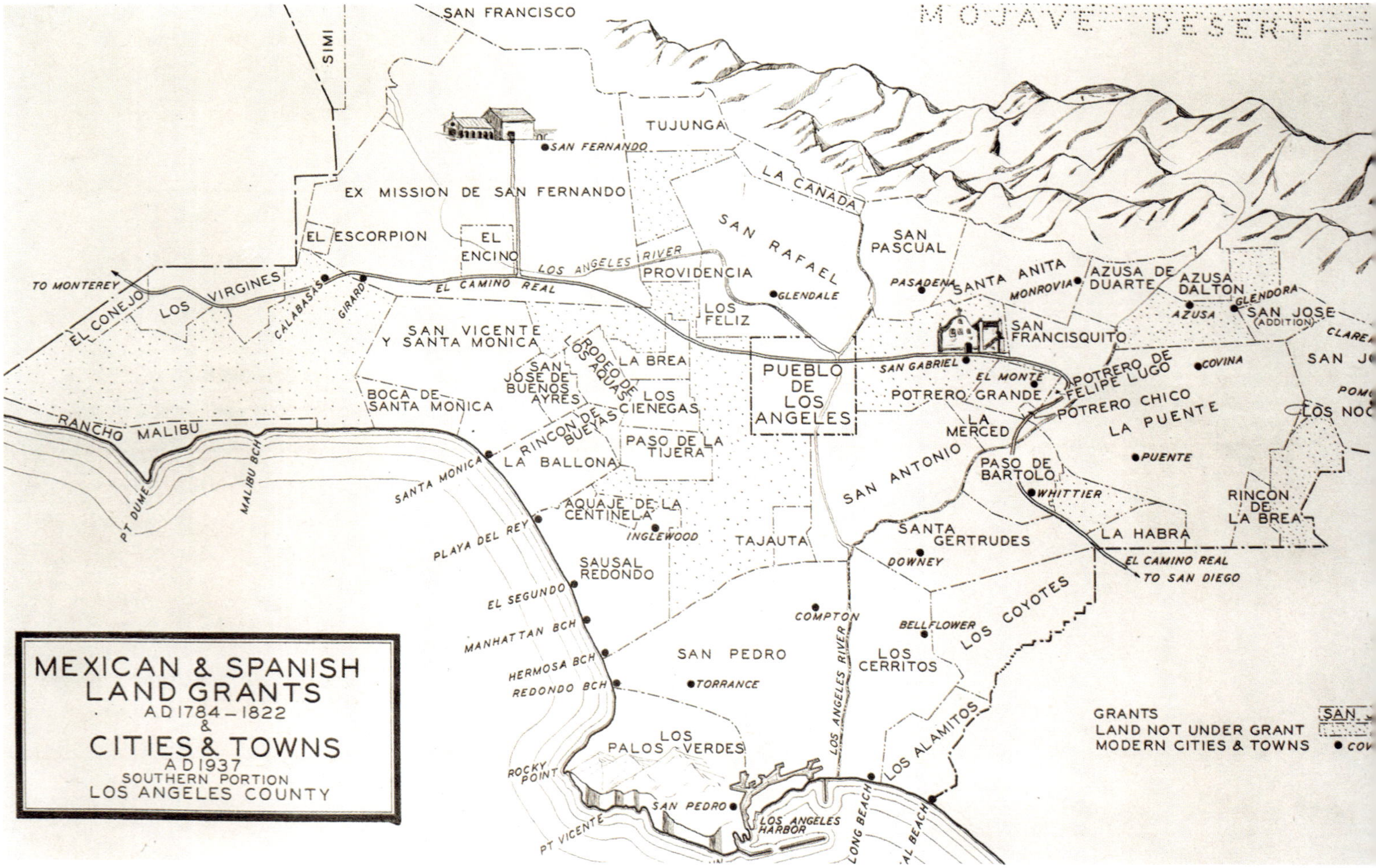

people and other newcomers to form a larger, if underappreciated, workforce. Native people from other parts of California came as well, drawn by work on the enormous *ranchos*. The hardest jobs were typically assigned to Native Americans. At the missions, priests doled out the jobs—from its earliest days, the mission system had a president-general (Padre Serra was the first) who determined what needed to be done. One president-general, Father Durán, griped, rather hypocritically, of *rancho* and pueblo residents: "If there is anything to be done, the Indian has to do it; if he fails to do it, nothing will be done. Is anything to be planted? The Indian must do it. Is the wheat to be harvested? Let the Indian come. Are adobes or tiles to be made, a house to be erected, a corral to be built, wood to be hauled, water to be brought for the kitchen? Let the Indian do it."

With their traditional hunting and gathering grounds impacted more every year by cattle grazing, the Gabrieleños and other native people had few options besides becoming neophytes at a mission, or working on a *rancho* or in the pueblo. On the *ranchos*, they were paid in food, clothing, and, sometimes alcohol—a practice that would become more common and destructive over the years. Although they worked hard for little benefit, Native American *rancho* workers were at least free from many of the restrictions that existed at the missions. They often preferred working on *ranchos*, cutting into the mission labor supply.

Competition for native workers was just one source of tension between *rancheros* and missionaries. An even bigger bone of contention was the land itself, with frequent disputes over boundaries and grazing land. *Rancheros* tried to seek land as far from the missions as possible to avoid such problems, since missionaries had political clout and usually prevailed, at least at first. Over the years, though, the power of the Franciscan missionaries declined as the *rancheros* began to gain influence. This was especially true after 1821, when Mexico became independent from Spain. Newly empowered, the *rancheros* and other *Californios* (Spanish-speaking Californians) wanted more land, especially the vast acres of farm and grazing land held by the missions. Ultimately, they helped to convince the Mexican government to take mission lands away from the church, a process that began in 1834.

With 1.5 million acres of land available from the Mission San Gabriel alone, the Mexican authorities made many more land grants, breaking up the huge properties formerly controlled by the Franciscans. Although a few Gabrieleños and other native people did receive land, for the most part, land originally intended for the neophytes was instead granted to soldiers, their families, and other private individuals. Often, more land was given to families who already had *ranchos,* increasing the wealth of a powerful few.

As local wealth increased with new access to land, new markets arose when *rancheros* and other residents wanted quality furniture, clothing, fine saddles, wine, brandy, and other luxuries. Mexico, finally independent from Spain, changed its laws in 1824 to allow for foreign immigration. Drawn by the expanding potential for commerce, Anglo businessmen and other newcomers began to arrive and settle in the Pueblo of Los Angeles in the 1820s and 1830s.

Abel Stearns typified the early Yankee immigrants to California who arrived eager for opportunity. An orphan from Massachusetts, he went to sea at twelve, and ended up in Mexico as a young man. Stearns eventually made his way to Los Angeles, where he found his niche as a middleman, buying hides and obtaining merchandise—such as fabric, clothing, dishes, sugar, tobacco, furniture, and wine—from trading ships. Reputed to be a smuggler, Stearns was often in trouble with the law, caught frequently in possession of contraband products such as liquor, silk, and improperly branded hides, suggesting that he was trafficking in stolen hides. After managing to evade not one but two California governors' orders of expulsion from the state, he went on to become a public official, holding several elected offices.

▲ Abel Stearns, one of the largest landholders in Southern California amassed over 200,000 acres. He built a warehouse at San Pedro Harbor to sell hides to trading ships and purchased goods for trading in clothing, fabric, furniture, dishes, and wine. Date unknown.

➤ Abel Stearns's *ranchos*, which covered part of modern-day Los Angeles and Orange Counties, were depicted in a lithograph that defined land ownership, town and section lines, railroads, drainage, and farming land. 1873.

Infamous as a smuggler, Stearns was also known for his appearance. Pueblo residents called him "Horse Face" due to his long, narrow features. His looks were not improved when a customer stabbed him in the face, angry at having purchased a barrel of spoiled wine. Despite his scarred, plain face, at age forty, Stearns courted and married fourteen-year-old beauty Arcadia Bandini, daughter of a prominent citizen. Once positioned as a member of a respectable family, Stearns was able to purchase Rancho Los Alamitos in 1842, a 28,000-acre chunk

MAP OF A PORTION OF
LOS ANGELES COUNTY Showing the
ABEL STEARNS' RANCHOS
5 Ranchos 200 square miles in blocks of 1 mile square
with road on each side. For sale in sections or fractions
by ALFRED ROBINSON Trustee at the
LOS ANGELES & SAN BERNARDINO LAND OFFICE
542 Market St. corner of Montgomery St. San Francisco
OR APPLY TO THE
AGENCY, LOS ANGELES OR ANAHEIM
EDITION MAY 1873
LITH. BY GEO. H. BAKER, 408 CALIFORNIA STREET SAN FRANCISCO CAL.
FARMING LANDS.
Early selections, under Grants from Mexico, in 1784 or 1834, confirmed by the United States Courts; Titles perfect. Distant six miles E. of the San Pedro and Los Angeles Railroad, now in operation; ten miles E. from Wilmington Port; Twelve miles S. E. from Los Angeles; 2 miles from Anaheim Landing, where the steamers regularly deliver and receive cargo.
The River New San Gabriel runs along the west side, the Santa Ana on the east, and the Pacific Ocean forms the southern front.
This large tract, nearly level, with a gentle slope southward towards the Ocean, comprises a variety of soil, rich and fertile, well adapted to wheat, barley, oats, potatoes, etc., and choice tracts for the culture of the lemon, orange, olive, fig, mulberry, grape, and tropical and northern fruits generally; also, cotton, hops, castor-oil beans, hemp, flax, and tobacco.
The large southern portion consists of bottom lands of rich alluvial and sandy loam, with water three to seven feet under and moistening the surface—especially appropriate for the cultivation of CORN on a large scale, without irrigation, from which land eighty to one hundred bushels per acre may be produced.
The famous German Settlement, "Anaheim," is on these lands.
For health and comfort the climate is excellent, as is well known.
Steamships leave San Francisco every six days for San Pedro, Wilmington and Anaheim Landing—running time, forty hours.
Freight from Anaheim Landing or vicinity to San Francisco, (by sailing vessels) $3 per ton.
The Government system of survey has been extended over these Ranchos for convenience and certainty of description.
TERMS—One-fourth cash; balance in one, two, and three years, with interest, payable at the end of each year, at ten per cent.
THE SHADED TRACTS ABOVE ARE SOLD.
PROPRIETORS.
E.F. NORTHAM,
EDW. MARTIN,
C.B. POLHEMUS,
EST. OF A. STEARNS,
GEO. H. HOWARD,
A.W. BOWMAN,
San Bernardino Base Line
LOS ANGELES
STEARNS' RANCHOS
SAN PASQUAL
SANTA ANITA
AZUSA
SAN FRANCISQUITO
SAN JOSE
EL MONTE
POTRERO GRANDE
Ro LA PUENTE
LA MERCED
PASO DE SAN BARTOLA
PASO DE LA TIJERA
RODEO DE LAS AGUAS
BALLONA
AGUAJE DE SENTINELA
FLORENCE
SAN ANTONIO Ro
SANTA GERTRUDES
GALLATIN
TAJUITA
SAUSAL REDONDA
COMPTON
DOWNEY
SAVANNA
SAN JUAN CAJON DE SANTA ANA
Kramer
LOS CIERRITOS Ro
LOS COYOTES
SAN PEDRO
LOS PALOS VERDES
WILMINGTON
LOS ALAMITOS Ro
ANAHEIM LANDING
LA BOLSA CHICA
ANAHEIM
CHAPMAN TRACT
SANTIAGO DE SANTA ANA Ro
SANTA ANA
TUSTIN
NEWPORT
SAN JOAQUIN RANCHO
PACIFIC OCEAN
T.1 S.R.14 W.
T.1 S.R.12 W.
T.1 S.R.11 W.
T.1 S.R.10 W.
T.1 S.R.9 W.
T.2 S.R.14 W.
T.2 S.R.13 W.
T.2 S.R.12 W.
T.2 S.R.11 W.
T.2 S.R.10 W.
T.2 S.R.9 W.
T.3 S.R.14 W.
T.3 S.R.13 W.
T.3 S.R.12 W.
T.3 S.R.11 W.
T.3 S.R.10 W.
T.3 S.R.9 W.
T.4 S.R.14 W.
T.4 S.R.13 W.
T.4 S.R.12 W.
T.4 S.R.11 W.
T.4 S.R.9 W.
T.5 S.R.14 W.
T.5 S.R.11 W.
T.5 S.R.10 W.
T.5 S.R.9 W.
T.6 S.R.9 W.

of the Rancho Los Nietos. As was typical of the time, Stearns paid its agreed-upon $5,934 price in hides and tallow, rather than cash. This purchase was just the beginning of his acquisitions. Stearns, once a penniless orphan, became one of the largest landowners in California, amassing more than 200,000 acres.

As more Anglos like Stearns became part of the local community, *rancheros* and newcomers with vested interest in the region's economic success grew frustrated with the politically unstable Mexican government. *Rancheros* and merchants alike wondered if Alta California would be better off as a sovereign nation, a British or French protectorate, or part of the United States. However, the United States' philosophy of Manifest Destiny envisioned expansion as a God-given right. The U.S. soon made clear that it intended to have California, even if it meant picking a fight. The resulting U.S.-Mexico War began in 1846. When it concluded with the signing of the Treaty of Guadalupe Hidalgo in 1848, Alta California became a territory of the United States. Another event, just days before the end of the war, would have equally profound implications: James Wilson Marshall, a carpenter who was constructing a sawmill for Swiss immigrant John Sutter, discovered gold at Sutter's Mill, changing California forever.

The Gold Rush created a long-lasting divide between Northern and Southern California. As emigrants flocked into Northern California, the north became wealthier and more metropolitan, while everything south of Monterey remained rural and sparsely populated, with little business beyond cattle ranching. *Rancho* culture intensified in all the counties from Monterey south to the Mexican border. These areas, viewed as unsophisticated backwaters by Northern Californians, were somewhat mockingly called the "cow counties." Los Angeles became known as the "Queen of the Cow Counties," perhaps because of the sheer size of its herds (or maybe because it was viewed as the most backward of all the southern counties).

Unrefined or not, Los Angeles soon became rich, thanks to its enormous herds of cattle. As would-be miners flooded Northern California, the price of beef skyrocketed, and meat surpassed the hide and tallow as the most lucrative part of the animal. While in years past, dried beef had been sold by the bale to trading ships, Los Angeles *rancheros* soon realized it was far more profitable to sell live cattle to the hungry forty-niners than it was to ship them beef jerky. Giant cattle drives from Los Angeles *ranchos* to San Francisco markets became a prominent feature of life.

The financial rewards were great when the *vaqueros* got the cattle to Northern California markets. Los Angeles was soon a wealthy town, bedecked with fine clothing, expensive furniture, and silver-trimmed

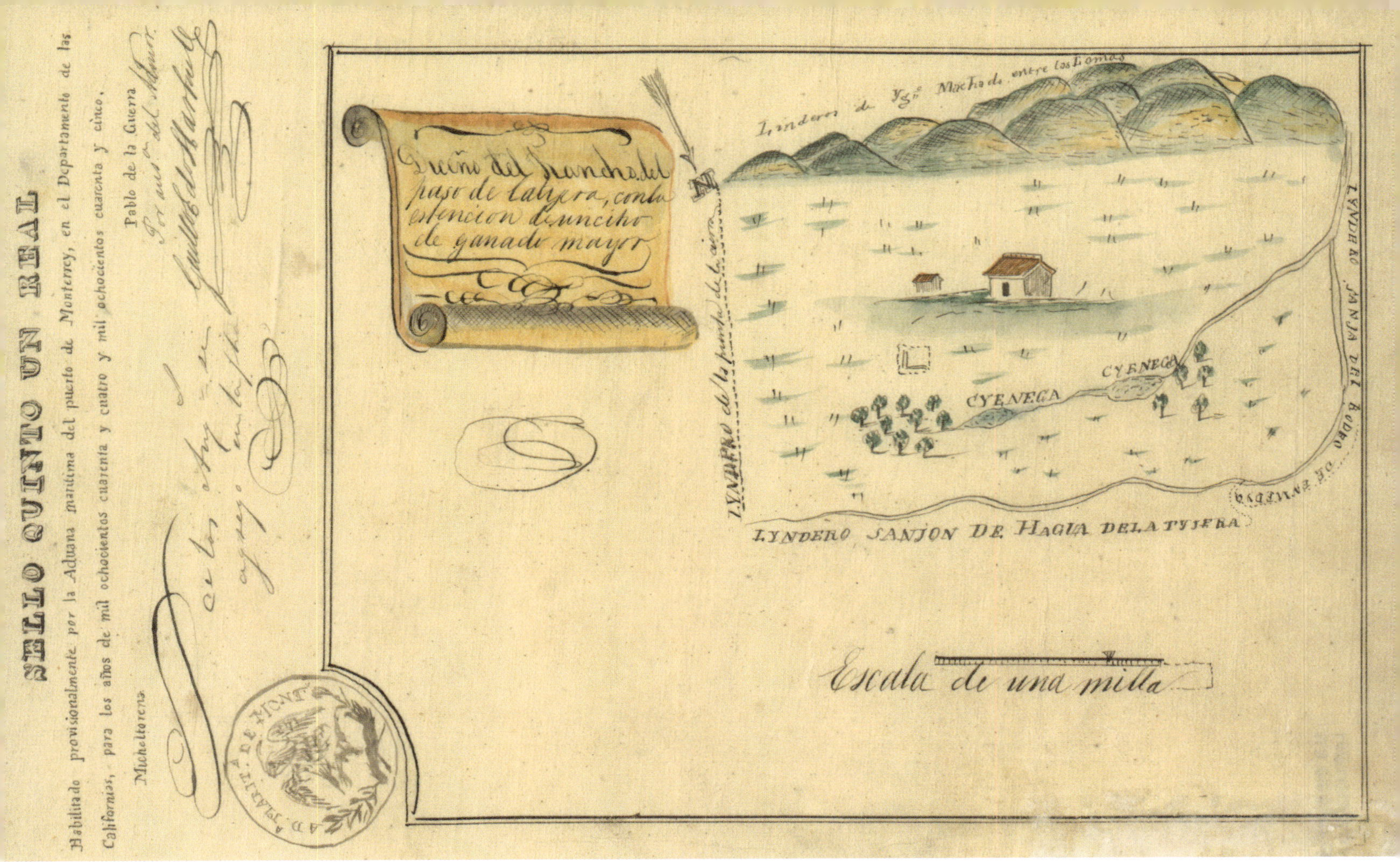

A *diseño*, or land case map, was an official record of land ownership in the nineteenth century. Historians use the maps as important research tools, especially for boundaries and landscape descriptions of early California-Mexican *ranchos*. Rancho Cienega o Paso de la Tijera includes the modern-day communities of Baldwin Hills, Windsor Hills, and much of Kenneth Hahn State Park. c. 1840.

saddles. "Everybody in Los Angeles seemed rich, everybody was rich, and money was more plentiful at that time, than in any other place of like size, I venture to say, in the world," wrote Horace Bell, an outspoken resident of early Los Angeles whose books about the region have become classics.

The bubble burst in 1857, as ranchers from Texas and Missouri began to drive cattle to market in San Francisco, overwhelming demand and driving down prices. Soon *rancheros* were back to slaughtering their cattle for hides and tallow, but now many were deeply in debt, having taken out loans to expand their operations.

Another threat to their way of life grew as well. The American public, inspired by the Gold Rush, now viewed California as the land of opportunity. Land-hungry settlers pushed west. Just as the *rancheros* had once viewed mission lands with a covetous eye, American settlers began to see *rancho* lands as ripe for the taking. Although the Treaty of Guadalupe Hidalgo guaranteed that Mexican land claims would be protected, Congress soon passed legislation that had grave implications for the *rancheros*.

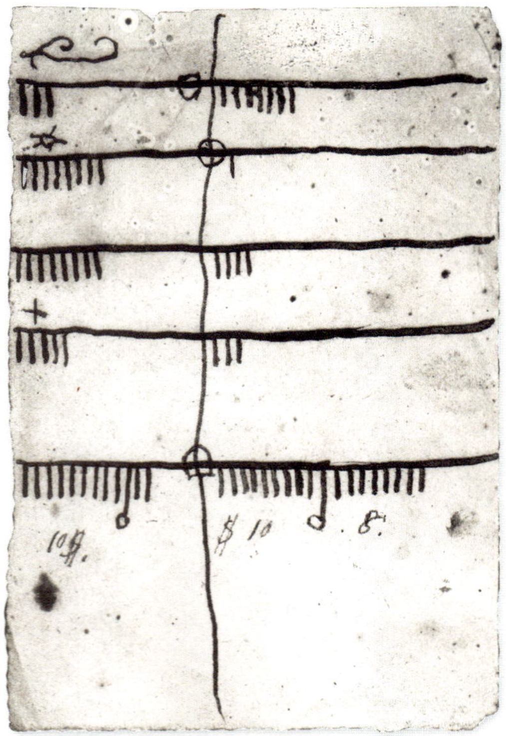

Rancheros used a tally sheet to calculate how many cattle were sold and the price of each. Date unknown.

Known as the California Land Act of 1851, the law created the Land Commission, a board to review all the land titles from Spanish and Mexican rule and determine their validity. This was a complicated matter, because during these eras, land ownership had been viewed very differently. Land had been plentiful, with cattle grazing on the open range, making precise boundaries irrelevant. Consequently, methods of defining property lines were inexact. Drawings called *diseños* showed general borders of *ranchos*, often using markers that were nonpermanent or difficult to find, such as streams, boulders, or even specific trees. This flexible way of thinking about property lines became a huge problem for *rancho* owners, who now had to prove—by U.S. legal standards—that they owned their land.

To make matters worse, the Land Commission met in San Francisco, a lengthy and expensive journey from Los Angeles in those times. *Rancheros* often sold off pieces of their property to raise the necessary funds for travel and legal fees. Lawyers often accepted land in lieu of cash, sometimes taking as much as half of a client's *rancho* to handle a claim. And land claims were lengthy, taking an average of seventeen years to resolve. *Rancheros* saw their wealth and land slowly erode as they went through the arduous land claim process.

Having endured so many setbacks, *rancheros* were hardly prepared for what would be the most crippling blow of all, a devastating two-year drought that began in 1862. Cattle dropped dead by the thousands, with the land between Los Angeles and Wilmington described as "a regular mass of dead cattle." *Rancheros* who had fetched up to seventy-five dollars per head for cattle during the height of the Gold Rush boom were now selling their herds for pennies per head, if they could sell them at all. More often than not, starving cattle were picked off by predators like grizzly bears—once common in Los Angeles—mountain lions, and coyotes, or simply died and rotted where they lay.

The drought ended cattle ranching as the central commercial venture in Los Angeles while setting the stage for future suburban sprawl. Over the decade of the 1860s, the number of cattle in Los Angeles County declined by more than seventy percent, from 70,000 head in 1860, to 20,000 or so by 1870. Los Angeles's reputation as "Queen of the Cow Counties" would soon fade away, but *rancho* culture left an indelible imprint on Los Angeles as we know it today. The prevalence of *ranchos* in the Los Angeles landscape was the precursor to rampant development, as private investors bought up *rancho* land for pennies an acre. Later, they would sell it for thousands after breaking it into smaller tracts.

The era of the open range was ending, and a new age of population and economic growth, driven by modern agricultural development, would take its place. Cattle ranching slowly became a relic. Today, it's rare to see cattle in the Los Angeles County landscape. At the Farm Sanctuary, Bruno the Steer seems like a novelty to visitors. On the tour that includes their visit with Bruno, visitors hear graphic details of how livestock suffer in modern corporate agriculture. Abel Stearns and other *rancheros* would no doubt find the Sanctuary, as well as the concept of animal rights, to be completely alien. They might relate better to another modern trend, the interest in grass-fed beef. Shoppers at upscale grocery stores and farmers markets around Los Angeles pay a premium for grass-raised beef, in part because they like the idea of cattle happily grazing in a green pasture. While they are unlikely to find beef raised in Los Angeles County, shoppers can find grass-fed beef from ranches in San Diego, Santa Barbara, and San Luis Obispo counties. In addition to its health advantages, which include lower fat and more vitamins, some proponents think grass-fed beef just plain tastes better.

Angelenos, like most Americans, just can't seem to give up their beef, despite a growing consensus in the health and medical community that eating too much red meat increases the risk of cancer and other chronic diseases. At the Farm Sanctuary, Bruno, sweet as a puppy dog, is safe and protected, a living representation of a bygone era, but most of his kind are destined for the dinner plate.

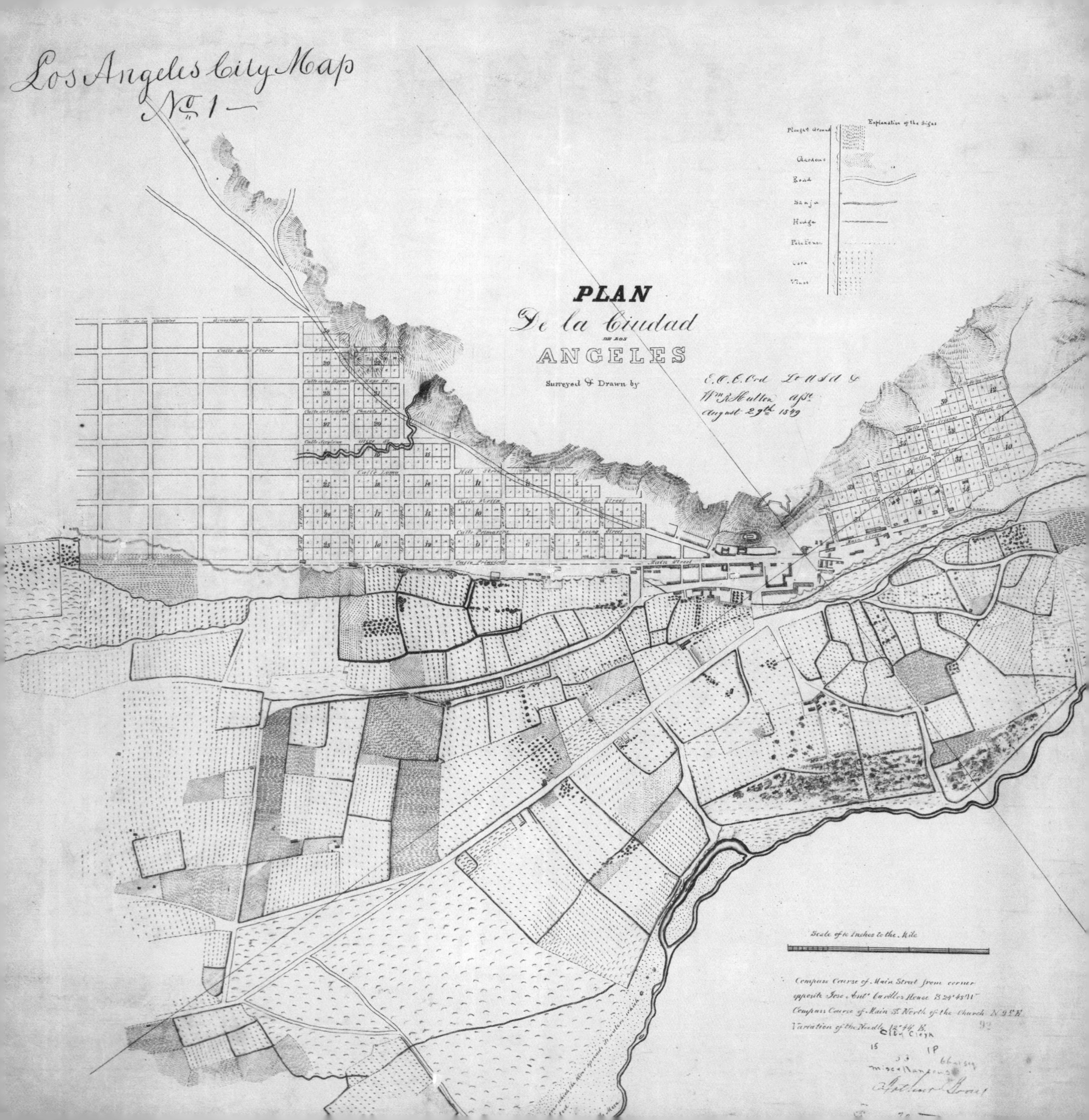

Los Angeles City Map
No. 1
PLAN
De la Ciudad
DE LOS
ANGELES
Surveyed & Drawn by
E. O. C. Ord Lt. U.S.A. &
Wm. R. Hutton Asst.
August 29th 1849
Explanation of the Signs
Plough'd Ground
Gardens
Road
Sanja
Hedge
Pole Fence
Corn
Vines
Calle de las Flores
Hill Street
Calle Mesita
Calle Primavera
Spring Street
Calle Principal
Main Street
Scale of 10 Inches to the Mile
Compass Course of Main Street from corner
opposite Jose Antº Carillos House S 24° 45' W
Compass Course of Main St. North of the Church N 9½° E
Variation of the Needle 12° 44' E.
City Clerk
miscellaneous

4 Early Grape and Citrus Farmers Transform Los Angeles

In 1854, the first Seal of the City of Los Angeles was adopted. Wine grapes were so important to early Los Angeles that a cluster of grapes surrounded by leaves is all that appeared. The seal was replaced on March 27, 1905, with another that depicts oranges, grapes, and olives, a reminder of the city's agricultural past.

A geologist visiting Los Angeles in 1853 wrote, "It was very delightful . . . to arrive in this vine-clad valley, and to walk through gardens and vineyards where the purple fruit hung in luxuriant and tempting clusters." Long before Sonoma or Napa became famous for their vines, Los Angeles was California's first wine country, leading the state's emerging industry for nearly a half century. Hemp, while it had been a promising crop, depended on subsidies from the Spanish colonial government, and fell by the wayside after 1810, leaving a continuing need for economic development. This need was filled by enterprising vineyardists, especially French immigrants. Vineyards were the first profitable business in Los Angeles besides cattle ranching, shaping the pueblo as it grew beyond its cow county roots.

The first Seal of the City of Los Angeles, the waxy stamp used by officials to approve public documents, adopted in 1854, showed how closely the young city identified with the wine-producing industry; it featured a cluster of grapes and grape leaves. The important road from the pueblo to the harbor in San Pedro was known as Vineyard Lane, and an early Los Angeles newspaper honored the grape and its resplendent vines by calling itself the *Southern California Vineyard*.

Plan de la Ciudad de Los Angeles, a survey map, part of the series done by Lieutenant Edward C. Ord. It provides detailed street and block information, but also shows the location of vineyards, cornfields, gardens, hedges, and fences. 1849.

As the state's skyrocketing population created a demand for fruit and wine, Los Angeles grape producers laid the foundation for another industry that would grow well into the twentieth century: they became California's first commercial citrus farmers. What started as a casual experiment by one Los Angeles grape grower, who planted orange trees out of curiosity, grew into a farming empire that left an indelible mark on Los Angeles. Once identified with its grape harvest, over time Los Angeles became associated with the golden fruit of orange trees. Today's massive California citrus and wine industries have deep roots in Los Angeles, having started out as cottage industries that boomed into professional enterprises during the Gold Rush. By 1905, the Seal of the City of Los Angeles would also include oranges.

Grape culture, from its promising start at the local missions, expanded into the Pueblo of Los Angeles long before the Gold Rush, with vineyards mentioned in public records as early as 1809. By 1818, there were some 53,000 vines established around the pueblo. Antonio Maria Lugo, a soldier and well known *ranchero* who served as mayor of the pueblo from 1816 to 1818, was likely the earliest vineyardist in Los Angeles, or at least the first whose name was recorded in official records. In fact, in 1818, Lugo, a corporal as well as mayor at the time, helped to capture Joseph "Pirate Joe" Chapman, a buccaneer who became another early Los Angeles grape grower. Chapman, a New Englander, was arrested at Santa Barbara as part of the crew of the Argentine privateer Hippolyte Bouchard, who carried out a plan to raid and plunder the Alta California capital of Monterey. Bouchard escaped with his ship and most of his crew after burning buildings and destroying property in both Monterey and Santa Barbara. Chapman, one of three captured crew members, was entrusted to Lugo, who brought his prisoner to Los Angeles, where he settled after spending time in jail. After learning the basics of viticulture while working as a handyman at Mission San Gabriel, Chapman left that job in 1926, purchased land in the pueblo, and planted four thousand vines using cuttings from the mission.

The grapes grown by early vineyardists were a Spanish variety known as the Mission grape, which thrived under the warm Los Angeles sun, but gave mixed results when fermented. Skilled mission vintners were able to use the Mission grape to produce quality wine, and Mission San Gabriel was especially

➤ Native Americans labored in the grape fields in the earliest days. Later, Chinese, Japanese, and Mexican workers harvested the crop. Sunny Slope Vineyard, San Gabriel Valley, c. 1900.

well known for a sweet, fortified white wine called Angelica. A century after the mission's Angelica was last produced, an English wine aficionado happened upon a bottle of this vintage, writing: "The wine was light brown in colour, rather syrupy, resembling a good sweet Malaga in taste, and in good condition." While some mission-produced wines were apparently drinkable for decades, Chapman and other less experienced winemakers produced wine that was "very inferior," in the words of a French captain visiting Los Angeles in 1827. The Mission grape, complained one vintner, produced a "strong wine . . . fit only for sailors to drink."

In 1858, Andrew A. Boyle became the first Anglo to settle east of the Los Angeles River where he bought land that was partly in the bluffs known as Paredon Blanco (White Bluffs), with the rest along the flatland next to the river. His land included a vineyard, and he bottled his wine under the Paredon Blanco name. Not only was he a successful vintner, he was also a member of the Los Angeles City Council in the 1860s. Date unknown.

French immigrants, led by Bordeaux native Jean-Louis Vignes, whose last name aptly means "vines" in French, improved the quality of local wine, setting the stage for Los Angeles to become a major wine-producing center. Like so many newcomers to early Los Angeles, Vignes came to reinvent himself. Already over fifty when he arrived in Los Angeles in 1833, Vignes had been a cooper, making barrels and buckets for a living in his homeland. Why he left France, leaving behind his wife and four children, is something of a mystery, although records hint that he may have been expelled due to his political loyalties. He first traveled to the Sandwich Islands (now Hawaii) hoping to establish a sugarcane plantation, but when his plans fell through, he headed for California.

Making his way to Los Angeles, Vignes purchased just over a hundred acres of land adjacent to the Los Angeles River, near the present location of Union Station. Here, he established his vineyard, El Aliso, and in less than a decade would dominate local winemaking. At the time Vignes arrived in Los Angeles, it was common to drink wine as soon as it was fermented. Vignes, known as "Don Luis" to the locals, became the first Los Angeles vintner to age his wine, using barrels he constructed with wood from local oak trees to produce smoother, more complex vintages. Dissatisfied with the quality of the Mission grape, he had cuttings of European varieties, including Cabernet and Sauvignon Blanc,

▲ Jean Luis Vignes named his winery El Aliso, for the large sycamore tree on the property that was once a meeting place for native people. Known as Don Luis del Aliso or simply "Don Luis," he was one of several French vintners who settled in the area near Union Station. Instrumental in bringing commercial viticulture to Los Angeles, Vignes was memorialized with a downtown street name. 1831.

shipped from France. By 1840, Vignes had over 40,000 thriving vines, producing enough to make regular wine and brandy shipments from San Pedro to ports in Santa Barbara, Monterey, and San Francisco.

Vignes influenced the demographics of the growing pueblo as he recruited his countrymen to make the long journey to Los Angeles. By 1850, one of every five Los Angeles residents was French or of French heritage, and locals often called the pueblo's vineyard district "French Town." Many of these immigrants became influential vintners in their own right, including Vignes's nephews Pierre and Jean-Louis Sainsevain, who became the makers of California's first sparkling wine, labeled as Sainsevain's Sparkling Champagne, first produced in the late 1850s.

Beyond his importance in Los Angeles, Vignes envisioned the enormous potential for wine production in California. According to California wine dealer William Davis, a close friend and confidante, Vignes told him that the day would come "when California would rival *la belle France* in wine producing of all varieties."

The Gold Rush spurred California winemaking toward this future, as thirsty forty-niners created a growing market for wine and brandy. Vineyards spread out along the banks of the Los Angeles River,

A view of East Los Angeles showing Brooklyn Heights (later known as Boyle Heights) in the foreground with the San Gabriel Mountains visible in the background. Early growers planted vineyards on the east side of the Los Angeles River, mostly in the area known as Paredon Blanco. One of these growers sold land to Andrew A. Boyle. 1876.

ROAD TO EL MONTE AND SAN GABRIEL

▲ The song "The Wines of Los Angeles County" was written by George G.W. Morgan for German vintners John Frohling and Charles Kohler, both influential in bringing large-scale wine production to Los Angeles County. 1876.

as local farmers expanded to meet the demand. Growers reaped great profits, in large part by paying native workers with alcohol to keep costs low. "The vineyards were cultivated by Indian labor. About all that it cost the vineyardists for labor was the amount of *aguardiente* that it took to give the Indian his regular Saturday night drunk. So the grape crop was all about profit," wrote J.M. Guinn in an early history of Southern California.

Pueblo officials facilitated the use of alcohol to secure and exploit Native American laborers, corralling intoxicated native people, then auctioning them off to the highest bidder for a week of labor. Farmers paid native workers in alcohol at the week's end, ensuring that the destructive cycle would continue.

The arrival of German immigrants Charles Kohler and John Frohling in 1854 further boosted local viticulture by creating a network to market Los Angeles wine. Kohler was a violinist who had settled in San Francisco, where he founded a classical music society. Frohling was a flutist in the society. The two friends were inspired to start a winery after eating some California grapes at a picnic, though they had no prior experience. They started with a fourteen-acre vineyard in Los

Angeles, expanding over several years into one of California's largest wine businesses, with a much larger vineyard in Los Angeles, a cellar in San Francisco, and sales offices in New York City and Boston.

When Los Angeles growers could not supply enough grapes to produce the quantities of wine needed for their growing empire, the enterprising duo orchestrated the 1857 founding of Anaheim, twenty-six miles to the south of Los Angeles proper. Best known today as the home of Disneyland, Anaheim was once a German colony organized to grow grapes and make wine. Although part of Orange County today, Anaheim was within the boundaries of Los Angeles County until 1889. Its prolific vineyards extended L.A.'s influence as the center of California's wine industry.

Although wine grapes helped to shape Los Angeles as it grew from a village into a small city, ultimately, no crop impacted Los Angeles as much as the orange. Successful wine grape growers became the first commercial citrus producers in California. Over time, many replaced their vineyards with the more lucrative orange and lemon groves. While many types of fruit would be cultivated in Los Angeles over the decades, oranges became the king of the orchard crops. Highly profitable due to growing demand for oranges and lemons, the vast citrus groves planted in the decades after 1870 brought riches to the growers and helped promote migration to Southern California. Orange trees, with their shiny green leaves and globes of golden fruit, became emblematic of Los Angeles, a symbol of its health and abundance that drew newcomers from around the country and the world.

The citrus industry, like the grape and wine business, was conceived in mission gardens. According to some sources, the mission padres did not readily share seeds or seedlings of their orange trees, so citrus did not spread much beyond the mission orchard until the church lost control of mission land in 1834. At that point, several Los Angeles viticulturists obtained seedlings from the mission and began to experiment with growing oranges.

William Wolfskill, a trapper and explorer born in Kentucky, was one of the first grape producers to grow oranges. He arrived in Los Angeles in 1831, soon planting a small vineyard which he expanded, becoming a leading grape grower in the pueblo. After admiring a few orange trees in the orchard of

John C. Wolfskill inspected oranges in the grove at the Wolfskill Ranch, located on Alameda Street, bordered by San Pedro and Ninth Streets, of what is modern-day downtown Los Angeles. c. 1890.

his neighbor, Jean-Louis Vignes, Wolfskill thought the fruit might be something he could sell. In 1841, to test his idea, he planted two acres next to his house with seedlings from the Mission San Gabriel orchard. This orchard, planted near Central Avenue and Fifth Street in modern-day downtown Los Angeles, was the first commercial orange grove in California. Within a decade, as his gamble on citrus paid off, Wolfskill expanded his grove to twenty-eight acres. The success of his orchard laid the foundation for what became a massive industry. Today, Wolfskill is sometimes called the "father" of California citrus farming.

Just as it had spurred huge markets for wine and beef, the Gold Rush also created a market for oranges and lemons as scurvy ravaged the miners. Once a plague among sailors and settlers, scurvy is caused by a Vitamin C deficiency brought on by an extended lack of fresh fruits and vegetables in the diet. Historical records suggest more than ten thousand people may have died of scurvy or related complications during the Gold Rush, since forty-niners had little access to fresh produce. While the science of vitamin deficiencies was not yet understood, it was known that eating fruits and vegetables, especially citrus, could help prevent or cure scurvy. Oranges and lemons commanded exorbitant prices as miners became desperate to protect themselves from the loose teeth, bleeding gums, diarrhea, and fatigue that were symptoms of the disease. San Francisco traders sent ships as far as Tahiti, Hawaii, and Mexico to secure lemons, oranges, limes, and other fresh produce, including grapes. Farmers in Los Angeles, much closer than these distant ports, stepped up to meet the demand.

As both grape and citrus farming expanded to meet the demands of the Gold Rush, water supply, a concern since the first colonists grew grain and hemp, became a more urgent problem. Farmers were limited by the reach of the *zanjas,* the open ditches that served as the main source of water for pueblo residents. Influential grape and citrus growers were instrumental in pushing for expansion and regulation of the *zanjas,* especially after Los Angeles became part of the United States in 1850. Under U.S. control, local officials more tightly regulated use of water from the *zanjas,* which were notoriously unsanitary. Used for bathing, washing, drinking (by both animals and people), as well as irrigation, contaminated *zanja* water frequently made people sick.

In 1854, the city created a new position, *zanjero,* to oversee its strained water system. The *zanjero* enforced rules, collected fines, and arrested people who committed crimes like dumping trash in the *zanjas* or stealing water. The *zanjero* was soon the most powerful official in Los Angeles, wielding more influence than the mayor, and drawing a larger salary as well. (William Mulholland would later become the best-known *zanjero.*) By 1870, fifty miles of *zanjas* crisscrossed Los Angeles. As the water system expanded, so too did vineyards and citrus groves.

Citrus farms multiplied even faster with the introduction of the Washington Navel Orange, a variety that soon eclipsed most others. Just as the quality of the Mission grape had been a limitation for early vintners, available orange varieties were a problem for citrus growers, as they were often bitter, dry, and full of seeds. In 1870, this was set to change when U.S. Department of Agriculture (USDA) scientists obtained a Brazilian orange that was sweet, seedless, and easy to peel. Known as the Washington navel, since it made its way from Brazil to California via Washington D.C., this orange arrived in Southern

The water wheel located near Elysian Park lifted water from the Los Angeles River into the Zanja Madre (mother ditch) and used wooden flumes to move the water to the brick reservoir in the Plaza. c. 1860.

California through the efforts of Eliza Lovell Tibbets, a resident of Riverside, sixty miles east of Los Angeles.

Among other things, Tibbets was an early advocate of women's rights, as well as a spiritualist who claimed to communicate with the dead (in fact, she met her third husband, Luther, when he hired her to contact his deceased wife). In 1873, fifty-year-old Eliza Tibbets emigrated from Washington, D.C., to the new colony of Riverside. Like so many new arrivals in Southern California, she and Luther decided to try farming despite a complete lack of prior experience, hoping to strike it rich with just the right crop.

In search of an appropriate farm venture, Tibbets wrote to an acquaintance, William Saunders, superintendent of USDA's experimental gardens in Washington, D.C. In 1873, Saunders shipped her two young specimens of the Brazilian orange trees that seemed promising. Tibbets planted the trees in her yard, where she watered them with her dishwater.

Since the fruit of the Washington navel orange is seedless, the trees could be propagated only by taking cuttings. With Eliza Tibbets' permission, neighbors used clippings of her trees to start their own. In fact, it was a neighbor who won first prize at Riverside's citrus fair in 1879, with fruit from a tree grown from a cutting. With this win came local fame, and the trees were soon in demand. For a time, the Tibbets' made a decent living selling cuttings from the parent trees for a dollar a piece, but since cuttings could be taken from any of the new trees, their business did not last. Eliza and Luther Tibbets would die penniless as others became rich growing the Washington navel.

➤ Upon the death of Andrew Boyle in 1871, his daughter and her husband William Henry Workman inherited his property, Paredon Blanco. A panoramic drawing shows the citrus orchard, vineyard, and Workman home, all of which were part of the inheritance. When Workman subdivided Paredon Blanco, he named the new neighborhood Boyle Heights to honor his father-in-law. 1880.

Food Forward volunteer Danielle Goldberg picks oranges in Woodland Hills. Thousands of acres of citrus groves once grew throughout Los Angeles County. Rather than let the fruit from orange and lemon trees go to waste, the non-profit Food Forward formed in 2009 to gather fruit from trees throughout the region and donates fruit to local food banks. 2014.

With the rise of citrus farming, orange groves were soon as prevalent as vineyards. As Los Angeles marked the centennial of its founding, vineyards and citrus orchards were its most notable features. James Miller Guinn described the scene in his *History of California and an Extended History of Los Angeles and Environs*: "From Main Street to the river, and below East Third Street to the city limits was a succession of orange groves and vineyards with an occasional walnut orchard interspersed. Looking down from the western hills, which then had a few scattered houses upon them, the observer beheld stretching away to the south for miles a sea of green."

From the 1880s into the early twentieth century, while citrus continued to ascend, grapes faded in importance as a local crop. A major factor in the decline of the Los Angeles wine grape industry was a mysterious disease that killed vines. Almost a century passed before scientists discovered its cause: a bacterium spread by insects such as leafhoppers. Called Pierce's Disease, it is still a problem for California grape growers today, though modern producers have more control options. Los Angeles area vineyards were more heavily impacted than those in Sonoma, Napa, and other parts of Northern California, in part because many Los Angeles viticulturists still grew the Mission grape, which was highly susceptible to the disease. Northern California farmers, on the other hand, tended to grow more disease-resistant European varieties. The northern part of the state also had a much larger population, with more wine drinkers. As a result, wine grape production shifted firmly northward to the counties around San Francisco. Many Los Angeles vintners, once hit with the disease, simply pulled out their sick vines, planting more lucrative orange groves instead.

The final blow to the local wine industry was Prohibition, which effectively ended Los Angeles's status as a significant wine producing area. Much of California's wine-grape industry was damaged, but not destroyed by the laws. For Los Angeles, though, with its wine grape production already severely in decline, Prohibition was the proverbial nail in the coffin. Today, just a single winery exists in the City of Los Angeles: the San Antonio Winery, begun by Italian immigrant Santo Cambianica in 1917, at a site just east of downtown L.A., a block from the Los Angeles River. San Antonio was one of the few post-Prohibition survivors among local wineries, thanks to Cambianica's savvy. He secured a contract with the Los Angeles Archdiocese for sacramental wine, a product exempt from Prohibition laws.

Despite the demise of the once-flourishing vineyards of Los Angeles, what began in the City of Angels positioned California for greatness, much as Jean-Louis Vignes had predicted. According to 2012 statistics, ninety percent of all wine produced in the U.S. comes from California, which follows France, Italy, and Spain as the world's fourth largest wine producer. An industry started in Los Angeles, at the local missions and in the tiny pueblo, spawned a massive business that dominates wine shops and grocery store aisles with products ranging from Two-Buck Chuck to celebrated vintages.

Citrus, as well, grew from its humble beginnings in Los Angeles. Today, California produces almost a third of the U.S. crop. While rarely grown for commerce in Los Angeles County these days, orange and lemon trees flourish in many suburban backyards. The amount of fruit generated by a few backyard trees is surprising, but not everyone harvests their crop, a fact noted by local resident Rick Nahmias. While walking his dog, Nahmias began to notice the amount of unpicked citrus fruit in the backyards of his San Fernando Valley neighborhood. He thought it would be a good idea to harvest some of this fruit for the needy, leading to a 2008 *Craigslist* posting for volunteer pickers. His idea spawned a thriving non-profit organization, Food Forward, which has donated millions of pounds of fruit to local food banks. Even with its bounty hidden in backyards, Los produces an abundant harvest.

➤ In 2016, San Antonio Winery, established in 1917, was the last remaining winery along the Los Angeles River and the last in the City of Los Angeles. It survived Prohibition by producing sacramental wines, like this Angelica, for the Los Angeles Archdiocese. Date unknown.

AVE. ELECTRIC RY
ST.
NINTH
CHANNING
ST.
LAWRENCE
ST
WARREN

5 Farming in the Time of the Railroad

On September 5, 1876, in a tiny, long-since-vanished Santa Clarita Valley town called Lang, forty-three miles north of Los Angeles, Southern Pacific Railroad president Charles Crocker hammered a golden spike into the tracks, marking the completion of the much-awaited line connecting Los Angeles and San Francisco. The mayors of San Francisco and Los Angeles were on hand to mark the occasion, as were hundreds of spectators and the thousand Chinese laborers who had blasted through a mile of solid rock in the Tehachapi Mountains to finish the line. A reporter from the *Sacramento Daily Union* captured the prevailing sentiment that the railroad would breathe new life into Los Angeles:

> The natural result of the driving of this last spike is that a tide of immigration must flow southward; that the lower counties of the State must shortly become as densely populated as the central; where hundreds of acres are now under cultivation, thousands will soon yield golden harvests.

The arrival of the Southern Pacific transformed Los Angeles, much as its supporters had hoped. With it came new opportunities for farmers, who could now ship perishable crops by rail to San Francisco and beyond. It brought thousands of newcomers, sparking an 1880s land boom that changed Los Angeles County from a rural outpost into something more like the sprawling patchwork of communities we

➤ Semi-Tropic Homestead Company commissioned a handpainted map to advertise its "117 fine lots for sale, $150 to $450 per lot, and $10 down, $10 per month, no interest." The Ninth Street Tract was located in downtown Los Angeles. Housing tracts were sold that offered farmland along with houses, often promoting images of the land's fertility. c 1880. ➤

know today. The opening of the railroad was a turning point for Los Angeles after two difficult decades following the collapse of the cattle industry and the catastrophic drought of 1862 to 1864. These in-between years were a time of experimentation, as Angelenos struggled with survival and hoped for success.

"Business is dull and times are exceedingly hard," wrote Henry D. Barrows, a vineyardist and son-in-law of famous citrus rancher William Wolfskill, in the spring of 1864. With demand created by the Gold Rush slowing, the future of farming in Los Angeles seemed limited at best, given its distance from markets

▲ The first California Fruit Train shipped fruit to the East Coast. The railroad opened vast new markets for Los Angeles citrus growers, and sparked a huge campaign to market both oranges and the Los Angeles sunshine. 1888.

and the scarcity of water beyond the network of *zanjas*. For a time, it appeared that Los Angeles was destined to be an economic backwater, with little potential for agriculture or industry beyond its orange groves, vineyards, and wineries.

Farmers tried out a myriad of crops, hoping to find anything that could turn a profit and thrive with limited water. From silkworms to cotton to pineapples, the farmers took on an array of farm ventures, and some were promising, but ultimately disappeared when they didn't pan out. Other non-starters included tobacco, sugar cane, peanuts, and castor beans. The reasons varied, but usually included the lack of local demand, since Los Angeles County was still rural and sparsely populated, with just over 11,000 residents in 1860. The population centers of Northern California and the Eastern U.S. offered more opportunity, but it was difficult and expensive to ship crops from isolated Los Angeles to these distant markets.

The few successes of the era included beekeeping and sheep ranching. Bee ranches, which dotted the foothills around Los Angeles, were popular in this post-cattle era, since they had low startup costs and no need for

This "future stretch of Wilshire Boulevard," as this photograph was captioned, became the site of the famed Ambassador Hotel. c. 1870.

irrigation. Large flocks of sheep also brought profits, especially after the Civil War disrupted availability of Southern cotton and increased demand for wool.

Wheat was the most important crop, by far, in the pre-railroad era. Because wheat is a water-thrifty crop, farmers could plant it when the first rain fell in the fall, and it could grow without further irrigation, making it possible to farm beyond the reach of the *zanjas*. And while it didn't bring huge profits, wheat was at least a fast crop, with a field of golden grain ready to harvest just a few months after planting. Like wool and honey, it was non-perishable and could be stored for long sea voyages. Much of the Los Angeles wheat harvest was shipped around the globe to Liverpool, England, then the major grain market for Europe.

Between 1868 and 1874, wheat production took off throughout Los Angeles County. The vast Rancho San Pedro, one of the few *ranchos* to remain in the hands of its original owners, stayed afloat because of large-scale grain farming. On the Rancho La Puente in eastern Los Angeles County, pioneer ranchers William Workman and John Rowland devoted thousands of acres to wheat. In the San Fernando Valley, partners Isaac Lankershim and Isaac Newton Van Nuys oversaw a massive wheat-farming enterprise.

Giant fields of wheat were the perfect temporary use of land, ideal for wealthy investors who had scooped up thousands of acres of former *ranchos*, then held on to the land as they awaited more lucrative opportunities. Those opportunities began to appear in 1876 with the arrival of the railroad, which radically changed the prospects of Los Angeles area farmers. With this new rail link to other cities, the long ocean voyage around South America's Cape Horn was no longer necessary to get crops to market in the Eastern U.S., creating new possibilities for shipping fruit and other perishable crops to distant consumers.

Perhaps more importantly, the railroad paved the way for large-scale migration to Southern California. What had been a steady stream of arrivals turned into a flood of people, creating surging demand for farm products as well as a red-hot market for land. Land owners quickly realized that subdividing their vast wheat fields into small farms and housing tracts would be the faster way to make money, not farming. As competition between the Southern Pacific Railroad and the newer Santa Fe Railroad helped drive down passenger fares, the cross-country trip was soon within reach for thousands of Americans. Real estate promoters were waiting for the visitors who came in droves, igniting a land boom that peaked in 1887.

Advertisements and travel literature created a popular mythology about Los Angeles as an agricultural paradise, inspiring thousands of would-be farmers to make the trip West. Charles Nordhoff's popular book *California for Health, Pleasure, and Residence: A Book for Travellers and Settlers,* published in 1872, had an especially strong influence on those with dreams of land in California. The Southern Pacific Railroad had commissioned Nordhoff, a well-known newspaperman, to travel California and then write a book that would encourage visitors to the Golden State. Though he had no particular expertise in agriculture, Nordhoff pronounced the superiority of Southland farms. "After a thorough examination of this region," he wrote, "I believe Southern California to be the finest part of the State, and the best region in the whole United States for farmers."

A man and his dogs stand in an almond orchard in the western Antelope Valley. Sixty miles north of downtown Los Angeles, the Antelope Valley, at the edge of the Mojave Desert, is part of Los Angeles County. In addition to grain and fruit, the Antelope Valley was a center for almond production at the time. According to the photo caption, this orchard was "the largest in the world." c. 1890.

Nordhoff's book, especially a chapter called "A January Day in Los Angeles," painted an idyllic picture of the Southland's pastoral abundance. He wrote:

> Around us the air was musical with the sweet smell of the baa-ing of young lambs. Surely, there is no prettier or kindlier sight in the world than a great flock of peaceful, full-fed ewes, with their lambs, covering a plain of soft green, as far as the eye can reach. Below us, as we looked off a hill-top, lay the suburbs of Los Angeles, green with the deep green of orange-groves, and golden to the nearer view with their abundant fruit. Twenty-one different kinds of flowers were blooming in the open air in a friend's garden in the town this January day; among them the tuberose, the Jessamine, and the fragrant stock or gillyflower.

Such depictions of Southern California's fields and flowers were often overblown and inaccurate. But accurate or not, visitors were often convinced that Los Angeles land was the holy grail of farming, much to the delight of local real estate developers. With carriage rides, free lunches, and brass bands, developers coaxed tourists to visit newly subdivided communities, where they would hear a sales pitch offering land on installment plans with low down payments. Realtors often touted the land's farming potential, like this advertisement for a subdivision that became part of today's distinctly unlovely and industrial City of Vernon:

> Catch on before the whole country rushes to Vernondale! Every man who wishes a home in Paradise should locate in this, the loveliest district of the whole of Southern California. This is where the orange groves are loveliest! This is where the grapes are most luxuriant!

Besides experimenting with silkworms, cotton, and even tobacco, some early Los Angeles farmers went for something even more exotic. J.B. Rapp's Hollywood pineapple plantation was the land that eventually became the northeast corner of Gower and Franklin streets. 1880.

As the boom spread throughout the county, more than sixty distinct towns and settlements "appeared like scenes conjured up by Aladdin's lamp. Out in the desert, in a river wash or a mud flat, upon a barren slope or hillside and in the fertile minds of the boomers these town sites appeared," according to land-boom historian Joseph Netz. Burbank, Glendale, Hollywood, Inglewood, Long Beach, Redondo Beach, San Pedro, and many more communities formed in the areas surrounding central Los Angeles during the boom. The

Workers harvest grain at the Lankershim Ranch in what is modern-day Van Nuys using mechanized farming equipment that sped up the process. The Ranch comprised 60,000 acres, with nearly 20,000 acres devoted to wheat in the late 1890s. The modern communities of Encino, North Hollywood, Sherman Oaks, Tarzana, Van Nuys, and Woodland Hills occupy former Lankershim Ranch land. 1898.

new subdivisions typically offered both residential and farm-size lots, ensuring continued growth of agriculture.

Water was a key factor of the boom, an essential for the multitude of new communities and farms to flourish. Beyond the reach of the *zanjas*, these communities relied instead on new sources of water developed as irrigation companies formed throughout the county. Water companies proliferated, obtaining rights to mountain springs to deliver the water to homes and farms in the valleys below, drilling artesian wells to tap underground aquifers, or diverting creeks with ditches and tunnels.

Access to water allowed farmers to transition away from wheat into higher-profit fruit and vegetable crops that required more labor, as well as more water. Increasingly, farmers fretted about the lack of farm workers. They could no longer rely on Native American workers, whose numbers were decimated by diseases such as smallpox and alcoholism, along with policies that destroyed Native American families. A state law passed in 1860, for example, allowed white people to hold native children as indentured servants until they turned twenty-five, making kidnapping of native children common throughout California.

The railroad, which solved so many other problems in the region, also brought a respite to the farm labor crisis in the form of Chinese railroad workers, who sought new employment as their construction work ended. Although a few Chinese had been in Los Angeles from the early 1850s, more arrived after the completion of the transcontinental railroad in Promontory, Utah, in 1869. Soon, Chinese men became the primary labor force for vineyards

Fields once plowed in the area of Vermont and Melrose in modern-day Hollywood eventually became the site of Los Angeles City College. This farmer was probably preparing to plant wheat, which was the most important crop around Los Angeles in the post-*rancho* era. 1874.

and citrus groves. Many had been highly skilled farmers in their homeland. Working in organized groups, they took work that no one else wanted: clearing land, digging irrigation canals, building fences, even acting as "human pesticides," killing insects by hand.

Chinese immigrants also contributed to the diversity and success of Los Angeles farming, becoming some of the earliest vegetable farmers in Los Angeles, growing leafy greens, cabbage, and green beans on vacant land near the *zanjas*. Building on their pioneering efforts, other farmers expanded into vegetable farming, especially as refrigerated rail cars became available. Green peas, cauliflower, and other vegetables could be grown in the warm Los Angeles winter, then shipped by rail and sold for a premium on the East Coast. Chinese farmers were often hired by Americans who wanted to start vegetable farming, but had no knowledge or experience. Chinese market gardeners, for example, helped white landowners develop celery farming on marshy land along the coast, helping to make celery an important and profitable crop.

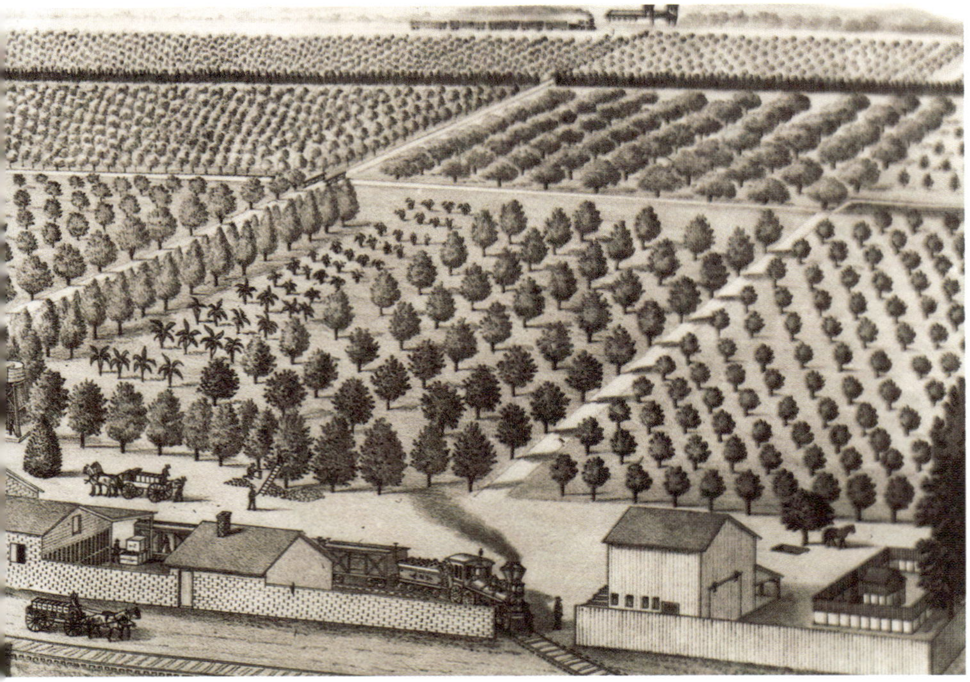

Citrus groves on the Wolfskill Ranch were located between Alameda and San Pedro Streets in downtown Los Angeles. William Wolfskill's orchard was the first commercial orange grove in California. The railroad made it possible for citrus farmers to get their fruit to markets far away. c. 1880.

Despite their contributions, Chinese Californians were often reviled by the larger population. A national economic depression in the 1870s sparked a long-lasting anti-Chinese movement, since many white Californians blamed their own lack of employment on the availability of cheap Chinese labor. Prejudice flared into violence. Some farmers

A panoramic image of Beachwood Canyon in Hollywood shows the Claussen Ranch and its orchards. Modern-day Gower Street runs up the hill starting near where the windmill once stood. c. 1890.

had to hire guards to protect Chinese workers from attacks by gangs of the unemployed intent on forcing the Chinese to abandon their jobs. Around Los Angeles, from the celery fields along the coast to the citrus packing sheds of the San Gabriel Valley, the Chinese were often harassed, beaten, and driven from their work. In 1882, anti-Chinese campaigns led to federal laws limiting Chinese immigration, eventually creating yet another farm-labor shortage.

Despite challenges with labor, the number of farms in Los Angeles multiplied during the boom years, expanding the production of fruits and vegetables throughout the county. Citrus ranching, which became ever more profitable with the advent of the railroad, continued its growth, especially in the foothills and valleys from Pasadena east to San Bernardino. Walnuts became another key crop, with major production in the San Fernando Valley, as well as the eastern L.A. County communities of El Monte, La Puente, Whittier, and the aptly named city of Walnut.

By early 1888, though, the land boom had reached a new level of frenzy, generating bad press around the country. "The Land Boom Run Mad: Waterless Deserts and Soilless Hills Selling at Fancy Prices" was the title of an article in the *New York Herald* in February 1888. Its writer painted an unflattering picture of Los Angeles, describing the place as a "dusty, straggling" town, with "abominable" food and innumerable, pesky real estate salesmen, a place where even grocers and tailors tried to sell visitors a piece of questionable property:

> The four or five principal streets swarm with real-estate men, and their placards line nearly every door post. The windows are full of them, boys distribute the brokers' "dodgers" on the sidewalk, sandwich men bear loud advertisements, and the very atmosphere seems charged with the all-pervading subject of local daily existence-viz.: Real estate to sell. How shall I get a customer?

The *Herald* reporter projected, like many pundits of the day, that the Los Angeles land bubble would soon burst. They were all correct. A few months later, by mid-1888, the land boom was a land bust.

Among those who had purchased small farms during the boom were many inexperienced newcomers who failed at farming and subsequently abandoned their land. Some had purchased land speculatively, with no intention of immediately farming or building, and many lost their money,

➤ A Sierra Madre Villa apiary overlooks the City of Pasadena. After the collapse of the cattle industry in the 1860s, "bee ranches" became popular. Unlike other crops, commercial honey production was possible with limited water access and a lack of farm capital. Beekeeping in Southern California was encouraged by John Muir, among others, who described California as "one sweet bee garden" due to its abundant wildflowers. 1888.

while others gave up on their investment and moved on. Newly subdivided land was neglected throughout the county, with farms unplanted and new homes empty. Some newly developed boom towns disappeared completely, including places called Gladstone, in part of what is now Azusa, and Kenilworth, which was carved out of a portion of the Rancho Los Felíz near Glendale. The population of the county fell month-by-month as newcomers pulled up stakes and moved on.

But while the momentum of Los Angeles agriculture had once again slowed, it had not stopped. The county's population, even as it declined, was still much larger than before the boom, growing from just over 11,000 people in 1880, to more than 50,000 in 1890. And despite the land bust, the railroad continued to thrive. With access to new markets via an expanding railroad system, sources of water beyond the city *zanjas*, and a wider variety of profitable crops, Los Angeles was poised for its next wave of growth. After 1890, its economy would still be centered on agriculture, but a new type of farming would evolve, one that was specialized, professional, and heavily focused on marketing. While many small farms failed, others became larger, focused on high-value crops that could be shipped across the nation by rail.

As farming became more professional and businesslike, though, the Angeleno fascination with real estate did not necessarily follow suit. *Los Angeles Times* reporter Thomas Curwen wrote in 2006, well over a century after the land boom ended:

> We are obsessed with real estate and always have been . . . Southern Californians have bought it, sold it, graded it, changed it, planted it, built on it, dreamed on it, torn it down and started all over again. We have pushed ourselves beyond our city centers. We have filled every space in between.

In Los Angeles, farming, real estate, and the railroad were inextricably connected. For decades after the land boom, the rise of professional farming would also drive the Angeleno passion for real estate. Images of golden oranges and prolific farms brought people from across the country and around the globe, with visions of owning their own piece of Southern California. The railroad made it possible, transporting these seekers to their imagined land of plenty, while delivering agricultural bounty to distant cities whose residents could enjoy its harvest, from sweet oranges to crunchy lettuce, even in the coldest winter.

◄ A Chinese worker gathers leaves at Curtis Silk Farms in Los Angeles. Curtis Silk Farms advertised their product in Sunset Magazine, where they offered a free "Beautiful Silk Book" with the purchase of a fifteen-cent skein of "beautiful golden silk." Silkworm culture or sericulture, was especially popular during the 1860s, when the California legislature had a program to promote silkworm culture in the hopes of developing a silk industry that would rival France, Italy, and China. 1907.

6 The Age of Professional Agriculture

Just twelve miles southeast of downtown L.A., close enough to the 605 Freeway to hear the rush of traffic, it's hard to imagine that fruit and nut orchards once dominated this strip-mall-lined stretch of Whittier Boulevard. But, just a few blocks away, sat the ranch home of farmer, feminist, and inventor Harriet Williams Russell Strong, the so-called "Walnut Queen" who helped make walnut farming a huge local industry. From 1890 to 1910, her efforts closely reflected advances in agriculture, when farming became a highly specialized business more like the corporate farming that evolved later in the twentieth century.

According to the U.S. Department of Agriculture (USDA), this part of Whittier is a food desert, a low-income area without enough affordable, healthy food nearby. It's a paradox that exists in much of Los Angeles County: communities where people have limited access to healthful food sit right on top of what was once rich farmland that grew an abundance of fresh produce. This particular area was part of the giant Rancho Los Nietos. Later, it was owned by Pio Pico, the last governor of California while it was still part of Mexico. His adobe home, on a tiny remaining portion of his ranch, is nearby at the Pio Pico State Historic Park.

◄ A man sits next to a peaceful Los Angeles River in Griffith Park. Many times, though, the river flooded, destroying crops and homes. As communities and farms spread out around Los Angeles, the river was no longer an adequate source of water. Many other sources were developed, including dams of mountain streams and artesian wells, but ultimately, larger scale solutions were pursued, leading to the construction of the Los Angeles Aqueduct, completed in 1913. The Los Angeles River was channelized in 1938, transforming much of the life-giving river into a cement-lined flood-control system. 1900.

In 1867, Pico sold a piece of his land to Harriet Strong and her husband, Charles Lyman Strong, who initially only built a home on the property. Charles, who worked in the mining industry, had little interest in agriculture, nor did Harriet, who had a painful spinal condition that often kept her

▲ Harriet Williams Russell Strong, a farmer, feminist and inventor, the "Walnut Queen" helped make walnut farming a huge L.A. industry. Date unknown.

confined to bed. But after her husband's suicide in 1883, the thirty-nine-year-old widow became determined to use their two hundred acres to support her four daughters, despite her own poor health and, perhaps worse, her complete lack of farming experience. She read books and articles, studying the science of farming, and also found that local farmers were happy to speak with her and give her candid appraisals of possible crops. As Strong noted,

> Had I been an Eastern man . . . I would have been told all sorts of tales . . . but they really gave me their best information. I had the courage of ignorance and plenty of determination to back it up.

Strong settled on walnuts as the best option to make a profit on her land, soon planting 150 acres of young walnut trees, as well as citrus, pomegranates, and other orchard crops. However, she would need to overcome two major impediments to her success: How would she support her family while waiting several years for her trees to reach maturity? And how would she water the trees on her arid farm? Walnuts could be grown without irrigation, but one dry year could mean disaster: losing trees. Strong could not afford that loss. Despite her chronic health problems, she pushed herself out of her sickbed, working tirelessly to solve her challenges. Her solutions went far beyond her own farm, making her a pioneer among Southern California farmers.

For a short-term crop, she struck upon pampas grass, a plant that produces fluffy white plumes, in demand for home décor among housewives of the era. After discovering that pampas grass was not only fast growing, but drought tolerant, Strong and her hired hands planted hundreds of plants in the rows between her walnut and pomegranate trees, making efficient use of her land and water. Pampas grass proved to be the salvation of her ranch as she waited for her walnut harvest. Within months, she was cutting and selling plumes, making 1,500 dollars from the first harvest. She became so successful at selling pampas grass around the country that by the early 1890s, she had orchestrated department store makeovers with the plumes, boosting their popularity nationwide. She even convinced both political parties to use the plumes, dyed red, white, and blue, to decorate their 1892 national conventions. A year later, the "Pampas Lady" (another moniker she picked up) took her plume show on the road to the 1893 World's Fair in Chicago, where she created a "Pampas Palace" from a million plumes.

Over the years, Strong's marketing efforts would continue, as she pitched and promoted not only her main crops of pampas grass

Young women who processed walnuts in a San Fernando Valley packing house work the lines. Date unknown.

and walnuts, but also pomegranates. As she wrote in an 1895 article for the *Overland Monthly*, a national magazine:

> The fruit of the pomegranate is first pleasant to the eye and good for food. Second, the juice is a refreshing drink for invalids, especially in fevers. Boiled, with the addition of sugar, we have a sirup for cakes and waffles that preserves a bewitching flavor.

Lack of water for her orchard was an even more daunting—and pressing—challenge, yet here she succeeded as well, becoming an inventor and leader in water issues. She designed a special type of dam that trapped and saved water in steep canyons, ultimately obtaining two patents for her invention. Taking advantage of a new state law, the Wright Act of 1887, which allowed the creation of irrigation districts that could levy taxes and issues bonds to build water infrastructure, she

Proudly displaying bags of sacked walnuts ready to be shipped from a walnut packing house in West Covina. 1905.

organized local farmers to construct the canals, dams, and reservoirs that would make it possible to coax a harvest from the dry land.

Strong was often in the spotlight since it was uncommon during this era to see high-profile women in agriculture. Those who lauded her were careful to reinforce her femininity, as demonstrated in this headline from an 1896 Chicago newspaper article: "She's Boss of Ranch. Yet Mrs. Strong is Dainty, Fragile, Little Woman. Her Hands Not Horny." However, this "fragile" woman—who was a vocal advocate for vocational training and voting rights for women and shared the stage with Susan B. Anthony on several occasions—focused on science, marketing, and infrastructure as the keys to successful farming.

Harriet Strong was right. After 1880, agriculture became highly specialized as farmers shifted away from wheat and into growing fruits, vegetables, flowers, and other crops that could be grown intensively on smaller plots of land. This new era was notable for its sophisticated marketing organizations, improved production through scientific research, and development of wholesale markets.

The Los Angeles Chamber of Commerce, founded in 1888 by business and community leaders, was instrumental in the transition of farming into an organized, sophisticated industry. Its members settled on farming as the solution to rebuilding the region's prosperity following the collapse of the land boom. Although the boom of the 1880s had ended, and many people departed as the boom became a bust, the population of Los Angeles County had still grown tremendously during the 1880s, up 40,000 residents by the end of the decade. This larger population drove up demand for food products. "Carloads of chickens, butter, cheese and eggs were pouring into the city every day, and heavy drafts were sent East and North in return," wrote author C.D. Willard in his 1899 history of the Chamber.

Chamber members believed in an opportunity for economic recovery based on meeting this demand for food, but only with a large population of "real" farmers—not the amateurs who had entered farming so full of hope during the boom years. With this in mind, the Chamber embarked on a massive campaign that would not only promote Los Angeles and its agricultural products, but also recruit experienced farmers, such as those from the Midwest.

To head up this campaign, the Chamber hired Frank Wiggins, a former hardware salesman from Indiana. Wiggins was one of the countless health-seekers who flocked to Los Angeles for its healing climate. In the nineteenth century, tuberculosis was epidemic in the U.S. The disease was usually fatal, causing

Frank Wiggins proudly poses with lemons. For nearly forty years, Wiggins had a profound influence on the development of the City of Los Angeles. His Chamber of Commerce campaign emphasized the "Land of Sunshine" and its leading industry: agriculture. His plan was so successful that millions flocked to L.A. to claim their own piece of agricultural paradise. Thus began the transformation from burgeoning town to international city. Date unknown.

those afflicted to slowly waste away while suffering from a persistent, bloody cough. Doctors believed that the only cure was rest and fresh air, if possible, in a warm, sunny climate. Thousands made their way to Southern California in hopes of regaining their health. While many such transplants continued to decline, dying far from home, Frank Wiggins was a success story. Suffering from tuberculosis, given just months to live, Wiggins not only recovered, but he became the quintessential booster of Los Angeles and its farm bounty.

Wiggins put together an elaborate, permanent display of Southern California products at the Chamber's headquarters, including a life-sized elephant made out of walnuts and a thirty-foot tower of wine bottles. He also helped to mastermind the massively popular California on Wheels exhibit, a rail display that traveled the country showcasing California products, such as farm goods from the Los Angeles area, attracting more than a million visitors. With Wiggins at the helm, the Chamber created special exhibits for expositions and fairs throughout the nation, such as a giant Liberty Bell constructed from Southern California oranges, and an enormous tower of oranges at the 1893 World's Fair in Chicago. Harriet Strong, who was by this time one of the first two female members of the Chamber, provided some of the oranges for the 1893 display.

More than selling Los Angeles farm goods, Wiggins and the Chamber sold Los Angeles to the nation, distributing millions of pamphlets that touted the county's superior qualities. "People are lectured, pelted, beaten, and bruised with an avalanche of information regarding Southern California, 'the garden spot of the world,' and 'the sanatorium of the universe,' a place to heal all their ills and troubles of every nature, kind, and description whatsoever," according to an *American Magazine* article

Jumbo the elephant, was made from 850 pounds of California walnuts and constructed for the 1893 Chicago World's Fair. The nutty pachyderm remained the chamber's star and was often taken on cross-country tours to showcase Los Angeles' agricultural bounty. c. 1925.

about Wiggins and this tidal wave of marketing hype. Over time, the campaign was successful in convincing many Midwesterners to try their hand at farming, Southern California style. The Chamber and its front man were so successful in promoting Los Angeles that city water czar William Mulholland later joked that the only way to slow the growth of the city would be to kill Frank Wiggins.

A young woman takes in the western view from Laughlin Park in Hollywood, a vista of plenty: undeveloped land and citrus orchards. 1905.

In addition to its relentless promotional campaigns, the Chamber helped to cultivate especially promising farm enterprises. In 1893, it encouraged citrus growers to organize into the Southern California Fruit Growers Exchange, later renamed Sunkist. A grower-owned cooperative, Sunkist was the premier example of the marketing associations that became instrumental to the success of farming, first in Los Angeles, then throughout California. In the coming years, producers of many other crops organized efforts modeled on its success, including the California Walnut Growers Association, organized in Los Angeles in 1912, which later became Diamond Walnuts.

Farmer cooperatives like Sunkist and Diamond were involved in every aspect of production, from picking, to packing, to distribution and marketing, in order to produce a uniform product that brought the highest price. U.C. Berkeley economist Paul Taylor wrote of the cooperatives that "even the work in the grove of pruning, fumigating, spraying, picking, and so forth, is commonly performed by gangs of laborers under contract, or employed and directed by the manager of the citrus association." As growers ceded control, becoming less hands-on, cooperatives fundamentally changed the way they interacted with their farms.

The cooperative movement also influenced the size of farms, pushing farmers to expand their acreage. While a farmer-owned cooperative sounds like an exercise in grassroots democracy, in fact, cooperatives often favored larger growers. While many citrus growers started out with small ten-acre groves, the cooperative movement pushed the trend toward larger acreage. Often, larger growers bought out smaller citrus farms, concentrating citrus production among fewer farmers.

By 1900, most Los Angeles-area citrus growers were more investors than farmers. As historian Carey McWilliams wrote in 1946, describing the wealth of these farmers: "to own an orange grove in Southern California is to live on the real gold coast of American agriculture. It is not by chance that the millionaire row in Pasadena should be called Orange Grove Avenue."

To bring the profits that these investor-farmers came to expect, cooperatives focused heavily on marketing. Sunkist, the premier example, was expert in selling citrus to an ever-expanding national customer base. Clever advertising campaigns touted the miraculous health benefits of oranges, which transformed the fruit from an occasional treat into a daily snack. Billboards, advertisements on streetcars, and displays of oranges in department store windows with slogans like "Fresh from California for Vigorous Health" and "A Big Help to Health" were hugely successful. In response to the growing demand, the number of orange trees throughout Los Angeles County expanded from less than a half-million fruit-producing trees in 1880, to more than two million in 1910.

⋎ Aside from citrus groves, the Hollywood area was also home to fields of flowers, like this sweet pea field near the area of Fairfax and Sunset that became West Hollywood. 1901.

As production of citrus, walnuts, and other high-value crops intensified, the demand for workers grew as well. Over time, farm jobs drew thousands of newcomers to the Los Angeles area, with a change in demographics every couple of decades: first Native Americans, then Chinese, then, beginning in the 1890s, a new immigrant group, the Japanese. Mexican, Punjabi Sikh, and white Americans also found work on Los Angeles County farms, but Japanese were the

Picking oranges on unstable ladders, while loaded down with burlap bags, was risky for workers, and injuries were common. Pickers had to follow strict guidelines for harvesting that not only made the process slow and laborious, but dangerous as well. Ranchers were adamant that fruit was not to be bruised during picking since damaged fruit could mold and spoil a shipment. 1895.

Grading oranges was an important process. A skilled grader could distinguish between different textures, shapes, and colors to determine quality. The best quality oranges were not only rated highest, but brought the highest prices. San Gabriel Valley. c. 1900.

Inset: Valley View Brand growers belonged to the College Heights Orange and Lemon Association in Claremont, another important area for citrus production. At one time Claremont had two packing houses along the Santa Fe rail tracks. Date unknown.

⮝ Chinese laborers worked on the Gates strawberry ranch in Eagle Rock. Due to exclusionary federal laws, while the Chinese had been the primary source of farm labor, during this era, Japanese laborers became increasingly important. c. 1900.

predominant farm labor force from 1890 to 1910, integral to agriculture in Los Angeles and throughout California. Before long, Japanese immigrants would become targets of harassment and exclusionary laws, like the Chinese before them. In the meantime, however, this new supply of cheap immigrant labor helped to fuel continued growth of farming around Los Angeles County.

Another factor in the commercialization of farming was the growth of a new industry: oil. Los Angeles was home to many natural asphalt reserves where tar seeped to the surface, the most famous of these being La Brea Tar Pits, now a tourist attraction. The Gabrieleños and other natives used natural asphalt for centuries to waterproof their boats and baskets. Later, the mission padres and pueblo residents used it to tar roofs. The tar was also a source of fuel for early Los Angeles, since it hardened into chunks that could be burned, though they were a poor substitute for coal. However, few Angelenos realized that in fact, tar seeps can mark the site of petroleum reserves, where crude oil has oozed to the surface, forming sticky tar.

It took an enterprising newcomer to make the connection between the local tar and a potential source of oil. In 1892, Edward L. Doheny, an unsuccessful silver prospector who had arrived in Los Angeles for a fresh start, noticed the local tar seeps. He wondered if they could represent underground oil deposits as was the case in other places, such as the McKittrick Tar Pits in Kern County, where prospectors were refining tar into kerosene.

Determined to explore the possibility that Los Angeles tar might represent a valuable commodity, Doheny and his business partner Charles Canfield decided to drill for oil at a tar seep near downtown. With a drilling rig fabricated from a eucalyptus tree trunk, they struck oil 155 feet down in late 1892.

Their makeshift well produced seven barrels a day, but this was only the beginning. It was not long before Doheny had more than sixty wells around Los Angeles. Within five years of the first strike, hundreds of wells owned by Doheny and others were pumping oil in the downtown area of Los Angeles. Fuel, which had been a limiting factor for industry, was now available without the high cost of transporting coal across the country. With affordable fuel came expanded production of manufactured food items such as canned fruits and vegetables, offering farmers more local options for selling their harvest.

The availability of oil also made irrigation possible for more farmers, since the gasoline-powered engine allowed farmers to pump water from deep in the ground. This was especially critical in the High Desert north of Los Angeles, which had suffered from two years of severe drought from 1895 to 1897, resulting in the failure of many farms, including one billed as the world's largest almond orchard. In 1904, when Antelope Valley farmers started using gasoline engines to pump water, farming in the High Desert resurfaced.

Alfalfa became an especially important irrigated crop, with fields of the bright green, clover-like plant seen from the High Desert to the coast. Produced for its high-quality hay, alfalfa fed the thriving Los Angeles dairy industry, which expanded to meet the local appetite for milk, cream, and cheese. Other farm industries boomed too; sugar beets, poultry, berries, olives, and flowers became increasingly important, as well as a seemingly endless array of vegetable and orchard crops.

Regardless of the crop they produced, Los Angeles-area farmers, like other farmers around the U.S., wanted research directed toward their needs. They clamored for improved crop varieties, better pest-control options, advanced machinery, and more ways to maximize their production. The value of scientific agriculture became

▼ A Japanese-American boy harvests carrots and loads them into a crate for the Rosemary Packing Corporation. Child labor was once extremely common in the fields around Los Angeles. Date unknown.

especially clear to Los Angeles farmers in the 1880s and '90s when the citrus industry faced a crisis related to a tiny pest. Cottony cushion scale, which feeds on the sap of citrus leaves and twigs, created a disaster for growers as infected trees began to decline and die.

"The cottony cushion scale would...seem to explode on trees, giving them the appearance of being caked in white cotton candy," wrote historian Douglas Sackman in his book *Orange Empire: California and the Fruits of Eden*. Farmers had few options other than to remove and burn their infested trees. For a time, it appeared that the pest might destroy the citrus industry. Ultimately, USDA scientists found a natural predator of the pest far away in its original home, Australia. First released in a Los Angeles orchard, Vedalia beetles, a type of ladybug, proved to be a voracious predator of the scale. The ladybugs saved the citrus business.

After a peach harvest in the San Fernando Valley, workers of all ages pose with the fruit of their labor. 1890.

Though pest control measures of the day were more likely to be focused on toxic chemicals like arsenic and cyanide than on biological controls, insects and diseases were a huge concern for agriculture. Farmers wanted scientific innovations on these and other topics, which pushed the University of California to expand its agricultural program at Berkeley to Southern California. In 1887, federal legislation provided the university with funds for agricultural experiment stations to conduct research on farm production issues. The initial field station was at Berkeley. Four additional stations were soon added around the state, with one established in 1890 in Pomona, on the eastern edge of Los Angeles County. The Pomona station conducted research on locally important crops, including sugar beets, grain, and grapes, but only planted two acres of citrus, a decision that would ultimately lead to its failure.

Citrus growers were unhappy with the Pomona station, both its location and its areas of research, and wanted a new U.C. experiment station that would focus exclusively on their needs. They successfully lobbied for a citrus research station, pushing a 1905 bill through the California Legislature to establish a station in Riverside. The beleaguered Pomona field station, suffering from lack of funding and local support, closed in 1907. In the coming decades, farmers would build on the political clout they flexed in securing the citrus research station, using it to develop powerful new advocacy groups like the Farm Bureau, which formed a decade later. This group in particular would play a key role in lobbying for programs and policies to support the county's agriculture, including more scientific research on local crops.

⮝ Baker Ranch pumpkins grew at the corner of modern-day Santa Monica Boulevard and Fairfax Avenue. First part of the ranch, in the mid-twentieth century, the intersection became known as the Crescent Junction District, a stop along the Pacific Electric Rail network. 1909.

Along with the growth of scientific farming, infrastructure developments such as the creation of centralized wholesale produce markets during this era benefited farmers. The city outgrew its early open-air produce markets around the Los Angeles Plaza by 1900, which were overwhelmed by the ever-larger volumes of fruits and vegetables coming in from outlying areas. The city needed a centralized system for handling farm produce, close to the railroad, where the region's bounty could be readily shipped to other cities. The first important wholesale produce market, known as the Third Street Market, opened around 1901, at Third Street and Central Avenue. In only a few years, the new market was so crowded with vendors that a group of Japanese immigrants started another market nearby, and soon created both competition and an expanded wholesale

produce area. Their Ninth Street Market, at the intersection of Ninth Street and San Pedro, opened in 1909.

Local pig farming got a boost from all the waste generated by the wholesale produce markets. "Just on the outskirts of Los Angeles, comprising a large part of the thriving community of Vernon, are to be found a great many pig farms, farms on which at least one-third to one-half of the fresh pork consumed in Los Angeles is raised," wrote the *Los Angeles Times* in 1909. "The owners of these farms are the chief scavengers on the market. They arrive long before rise of sun in the morning and move from stall to stall, selecting box after box of peaches, pears, tomatoes, melons, or other products which have become soft."

Men and women pick strawberries in a location only identified as near Los Angeles. The South Bay communities became well-known for their berries, including blackberries, raspberries, and especially strawberries. One community, Gardena, became known as the "Berry Capital of Southern California," and "Berryland." Date unknown.

By 1910, farming in Los Angeles, which began as a haphazard experiment, was well established as a professional industry. Los Angeles County's agriculture ranked not only first in California, but also number one among counties in the United States. State agricultural statistics from 1910 reveal the scope and diversity of the county's nearly eight thousand farms. The harvest was lucrative, with the total value of all Los Angeles County crops approaching fifteen million dollars.

A snapshot of the top crops for 1910 showed that L.A. County led the state in fruit and nut crops, the number one producer of walnuts, lemons, and strawberries, and among the top counties in oranges, almonds, olives, and apples. Vegetable crops were important as well, with Los Angeles leading the state in its harvests of cauliflower, cabbage, lettuce, and tomatoes. Los Angeles County was the top producer of corn, as well as hay, thanks to productive alfalfa fields. Animal agriculture was booming, too, with more than 20,000; cows, and a half-million chickens on farms and ranches around the county.

Sorting olives at the Los Angeles Olive Company. Date unknown.

The first important Los Angeles wholesale produce market opened around 1901, at 3rd Street and Central Avenue, and was known as the 3rd Street Market. The market was one of many infrastructure developments making it easier for farmers to market their produce. c. 1900.

Sheep were herded near modern-day Washington Street and Adams Boulevard. Later, Culver City would be home to film studio backlots. c. 1900.

More than half of the county's farms were irrigated by 1910, a key driver for the diversity and abundance of local agriculture. Harriet Strong, who had been one of the early adopters of irrigation, would go on to advocate for water development in the Western U.S., speaking before Congress on the importance of dams and reservoirs that expanded irrigated agriculture. Through such efforts, she influenced major public works projects that would bring water to the West on a massive scale, including the All-American Canal, which delivers Colorado River water to California's Imperial Valley. Strong died in a 1926 car crash before she could see some of her ideas come to fruition in the passage of the 1928 Boulder Canyon Project Act that created the Hoover Dam, a massive public works project that would provide water for farms in Arizona, California, and Nevada.

H.G. Keisling was awarded an official first-prize certificate for a Mottled Langshan Hen at the Los Angeles County Poultry Association show. 1892.

A woman with influence locally and on the national scene, Harriet Strong's impact on farming has largely been forgotten. Nonetheless, she and others of her era left an enduring legacy. The continued growth of marketing cooperatives, farm research, irrigation, and market infrastructure would allow Los Angeles to hold its position as the nation's top farm county for four decades.

In 1872, the Southern District Agricultural Society purchased 160 acres to use as a fairground and exhibit space for area agriculture. Located at modern-day Figueroa Street and Exposition Boulevard, Agricultural Park, which would become known as Exposition Park, had its entrance at Wesley Avenue. Date unknown.

The crate label for Sunkist Cooperative member Mission Play Brand oranges romanticized the mission era, and was used to promote Southern California using stereotypical portrayals of both Native Americans and Mexicans. 1923.

7 Fantastic Farms and Gardens

"From 1900 to 1920, Los Angeles was essentially a tourist town," wrote historian Carey McWilliams. "During the winter months, Los Angeles, was, in fact, a great circus without a tent." Beginning in the 1880s, thousands flocked to the area to enjoy the sunshine, see its bountiful farms and gardens, and experience its somewhat eccentric attractions.

During the same years that commercial farming was developing around Los Angeles, so too was an industry that was part farm, part sideshow. The *Los Angeles Times* described these enterprises as "fantastic farms," businesses that developed "some exclusive breed of livestock for commercial purposes—some type of beast, bird, or good red herring so strange that the public clamors to look upon it and marvel. And the more unusual the stock happens to be, the more callers it is likely to entertain." Ornate gardens, as well, were a huge draw for visitors. The local passion for ornamental plants spawned a nursery industry that would eclipse other types of farming within a few decades.

Ostrich farms were the earliest and most popular of these farm attractions, drawing tourists by the thousands. Though ostrich farming may sound whimsical, it actually had the makings of a highly lucrative enterprise, driven by a global demand for feathers to decorate the elaborate women's hats of the day. From the 1880s until World War I, ostrich feathers were so valued as a commodity that historians and economists have compared the international trade in plumes to the diamond trade. In fact, in the early twentieth century, a pound of ostrich feathers was worth almost as much as a pound of diamonds. Ostrich farming originated

A man taught a group of school children about ostriches at the Los Angeles Ostrich Farm, Lincoln Park. The farm claimed to have the "only trained ostriches in the entire West," but in fact, the birds had a habit of kicking when agitated. To avoid dangerous outbursts, handlers blindfolded the ostriches who then couldn't see potential threats. Date unknown.

The Cawston Ostrich Farm in South Pasadena was one of two popular ostrich farms in the Los Angeles area, and the biggest, offering tourists the chance to take ostrich-drawn carriage rides or ride on the back of an ostrich. Considered one of the region's first "amusement parks," it survived for nearly forty years, operating from 1896 to 1935. 1916.

in South Africa, where it made millionaires of farmers and merchants alike. The enterprise seemed like a natural fit for Southern California, which features a similar climate.

At first, Southern California ostrich farming was strictly about production, not tourism, perhaps due to the unpleasant disposition of the animals, problematic birds that are nervous and prone to stampeding when startled. "An ostrich is apparently about the most ill-tempered bird in existence. They never acquire a fondness for anyone. They are always on the lookout to kick someone, and if the kick has the intended effect, it is pretty sure to be fatal," according to a *New York Times* reporter who visited Southern California's first ostrich farm near Anaheim, owned by Englishman Dr. Charles J. Sketchley. Before long, though, it became clear that ostriches were a potentially lucrative tourist attraction. After ongoing problems with curious trespassers, including some who attempted to illicitly (and unwisely) pluck feathers from the birds, Sketchley concluded that visitors might as well be supervised and charged admission.

He partnered with Welsh-born journalist and mining investor-turned-rancher Griffith J. Griffith to open Los Angeles's first ostrich farm designed with visitors in mind. This farm was located on the same chunk of hilly property that was once part of Rancho Los Feliz. Eventually known as Griffith Park, these four thousand acres eventually housed the L.A. Zoo and attractions such as the Greek Theatre and the Griffith Observatory.

Established in 1885, the ostrich farm was the first tourist attraction on this land, complete with its own "Ostrich Farm Railway" to bring tourists from downtown L.A., allowing the curious to spend the day

▲ A Cawston "Girl" wearing an elaborate hat adorned with one of the famed Cawston Ostrich Farm's feathers. In the late nineteenth and early twentieth centuries, ostrich feathers were a popular fashion accessory used to make boas, fans, muffs, and to add a dramatic flair to the large hats popular at the time. c. 1900.

viewing the ostriches while picnicking among the oak trees. This early Los Angeles ostrich farm lasted only four years, but was followed by others, all of which attracted throngs of tourists.

The most famous, by far, was the Cawston Ostrich Farm in South Pasadena, which opened in 1896. Visitors watched ostriches swallow whole oranges, rode in ostrich-drawn carts, and had their pictures taken sitting on the gangly-but-strong birds. The farm also included an on-site feather-processing plant along with a showroom that featured thousands of plumes for sale. At the height of the ostrich-feather craze, Cawston's had a retail store at 313 Broadway in downtown Los Angeles, as well as stores in San Francisco, Chicago, and New York.

The Los Angeles Ostrich Farm in Lincoln Heights neighbored the California Alligator Farm that opened in 1907 as a one-stop shop where, according to brochures, visitors could see "hundreds of alligators of all sizes, from little babies, hardly the size of a lizard, up to huge monsters, five hundred years old or more." The brochure also advertised fashion accessories made with alligator skin, including their specialty, "alligator bags ornamented with genuine alligator heads and claws."

While the California Alligator Farm was gated to keep the reptiles away from the community at large, in what seems to be a bit of irony, the gates did not keep visitors safely away from the gators. In fact, visitors were encouraged to interact with them by "riding" a saddled Billy the Alligator, one of the farm's famous tenants who made frequent appearances in motion pictures. They also walked and swam alongside the gators in the farm's pens and miniature lakes. Occasionally, unfortunate accidents did happen with alligators in such close proximity to people. In one instance, a gator jumped up and grabbed a seven-year-old boy's hand in its jaws, when the child climbed a fence and got too close at feeding time. According to the *Los Angeles Times*, the child's quick-thinking uncle saved the day when he "gouged the alligator's eyes, thus forcing him to open his jaws." In another incident, a gator bit off a keeper's finger at feeding time. "Finger is Added to an Alligator's Breakfast Food," was the title of the resulting article in the *Times*.

Ostrich and alligator farms were just the beginning of the animal-themed farm attractions around Los Angeles. From a monkey farm in Culver City, to a snake farm in North Hollywood, to the famous Gay's Lion Farm in El Monte, animal farms proliferated. "Of the goldfish, canary, and pheasant farms, their number is legion," wrote the *Los Angeles Times*.

Tropical greenery was also a feature at the "fantastic farms," adding to their exotic appeal for tourists. Gardens were part of the Cawston's experience, for example, which their brochure described as "a semi-tropical ever-green Ostrich Paradise, where people gather to enjoy the beauty of California flowers, plants, and palms, and to be entertained by the strange, immense and extremely foolish birds that live in the luxury of greenery instead of the desert wilds of their native land."

Elaborate gardens, like ostriches and alligators, were a big draw for travelers in the Southland. The extravagant home and garden of French painter Paul de Longpre was one of Hollywood's first attractions, a popular stop for tourists. Known for his botanical paintings, de Longpre created a stunning three-acre garden, featuring hundreds of rose varieties, around his mansion at the corner of Hollywood and

The California Alligator Farm at Lincoln Park was located next to the Los Angeles Ostrich Farm. Apparently Angelenos in the early 1900s didn't think it was dangerous to allow their children to ride an alligator, since the farm had "competent" guides who "trained" and "hypnotized" the alligators. The brochure boasted that the farm had more than a thousand gators, making it the world's largest alligator farm. Date unknown.

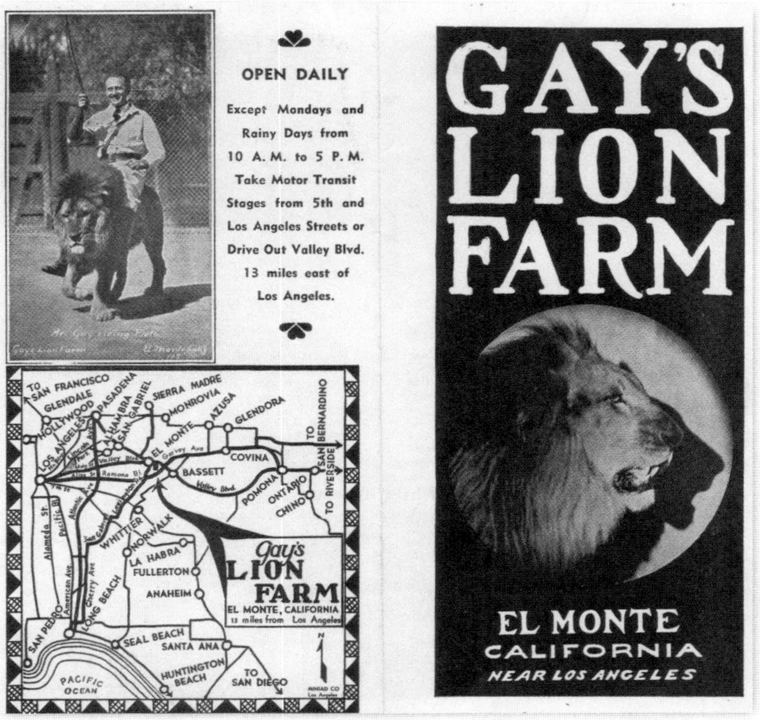

⮝ Lion "farms" were added to the mix of farm oddities in the early twentieth century. Charles Gay was a lion trainer who had trained lions for circuses, zoos, and most famously, for the motion picture industry in MacArthur Park (then known as Westlake Park). His most famous resident served as the face of MGM movies. As his popularity grew, Gay relocated to El Monte and opened up his farm as a tourist attraction. Date unknown.

Cahuenga Boulevards. In 1910, director D.W. Griffith filmed *Love Among the Roses*, one of the first motion pictures made in Hollywood, which starred a teenaged Mary Pickford, in the de Longpre Garden. At the peak of its popularity, eight thousand people toured the garden each month.

Flower farms also drew tourists, who flocked to E.J. Vawter's carnation fields in the Ocean Park neighborhood of Santa Monica, as well as the many acres of fragrant sweet pea fields in Gardena. Redondo Beach, sometimes called "Carnation City," was another stop for tourists, who visited its twelve-acre carnation farm that produced as many as 10,000 blooms a day.

From ostrich farms to lavish gardens, tourist attractions catered to the traveler's passion for the exotic. The Midwesterners and East Coast residents who flooded Los Angeles were part of a growing American middle class who had had little opportunity to travel. They craved something foreign and exciting, without the trouble of crossing an ocean or a border. Extravagant blooms, tropical foliage, and strange animals were part of the payoff for making the long journey across the country. Of course, Los Angeles boosters hoped that many of the visitors would stay. Historian Remi Nadeau wrote: "Tourists and settlers were, and still are, regarded as an annual crop, like oranges. With them, they brought the savings of a lifetime." By playing up the exotic, promoting Los Angeles as the "Riviera of the nation," "American Italy," "Land of Sunshine," and an array of other monikers, boosters ensured that thousands would come, and many would stay.

Once they settled in Southern California, many new residents created their own gardens, seeking

out trees and plants that would transform the landscape into their own semi-tropical paradise. For the wealthy, in particular, elaborate gardens represented success. “A man’s status was symbolized by the sweep of the lawn that separated his house from the street and the number of specimen trees and palms that grew thereon,” wrote Southern California garden historian Victoria Padilla in 1961.

In the citrus communities of the San Gabriel Valley, especially, residents newly rich from oranges, real estate, or some other aspect of the booming economy created elaborate gardens as an expression of their wealth. One notable example was railway magnate Henry Huntington, who in 1905 hired German horticulturist William Hertrich to develop a showcase garden on

▲ Artist Paul de Longpre and his daughter pick roses from one of four thousand bushes in his rose garden, part of a three-acre garden on his sprawling estate. De Longpre’s garden was one of Hollywood’s first tourist attractions. Each month, as many as eight thousand visitors came to see hundreds of rose varieties and tour the opulent house. One of the first motion pictures made in Hollywood, *Love Among the Roses*, directed by D.W. Griffith and starring fifteen-year old Mary Pickford, was filmed in the de Longpre Gardens. 1910.

➤ A fashionable young woman seems to be thinking about feeding one of her delicious Sunkist oranges to a lion at Gay’s Lion Farm. Date unknown.

Sunkist Oranges
CALIFORNIA
EXCHANGE

his ranch in San Marino. Hertrich used a horse and wagon to bring in large specimen trees, especially palms. He traveled the world to find plants for Huntington, visiting New Zealand, Fiji, and the Caribbean in his quest to obtain exotic additions for Huntington's garden, which became the Huntington Botanical Gardens after Huntington's death in 1928. The elaborate botanical park that Hertrich created, which included a desert garden, a palm garden, and a Japanese garden, evolved and expanded over the years into a world-class collection featuring more than 15,000 plant varieties.

The exclusive community of Chester Place, south of downtown Los Angeles was home to oil tycoon Edward L. Doheny, who was also a major plant collector. Doheny kept an ever-expanding collection of palms in a large conservancy at his Chester Place home. He sent nurserymen to collect exotic plants, especially palms and cycads, to locations including Mexico, Cuba, and Guatemala. Date unknown.

Wealthy amateur botanists like Huntington and oil magnate Edward Doheny, another plant collector, introduced plants from far-flung locations around the world to Southern California. Plants native to Australia and South Africa became especially prevalent in local gardens, as did palms, which came from many places, including the Sonoran desert of Mexico and the jungles of Central America. Doheny sent nurseryman Edward Howard to seek out palms in Mexico and Guatemala, which Howard hauled back by train to showcase at Doheny's mansion on Chester Place, an exclusive street south of Downtown. Palms introduced by Huntington, Doheny, and others brought the feel of the tropics to the streets of Los Angeles and became iconic features of the local landscape.

Average Angelenos were passionate gardeners, too, especially the many transplants from colder climates. "Even the most unpretentious dwellings were surrounded by profuse plantings of every hue, fragrance, and form," according to Padilla. "Many newcomers to Southern California, excited at finding

Theodore Payne's first store on South Broadway in downtown Los Angeles. Theodore Payne worked for the Germain Seed Company before leaving to start his own business focusing on native plants. Payne helped to design native plant gardens at Caltech, Descanso Gardens, Exposition Park, and the Rancho Santa Ana Botanic Garden. The Theodore Payne Foundation for Wildflowers and Native Plants in Sun Valley, founded in 1963, carries on his legacy of educating the public about native plants. c. 1910.

themselves in a land of no snow and but little frost, liked to believe that they were actually dwelling in the tropics and grasped at everything that would make their gardens suggestive of the more torrid zones. Geraniums, cannas, poinsettias, bougainvilleas, bananas—all those plants which are tenderly nurtured under protection in the East—were grown in abundance."

Many small plant nurseries started up to meet the ever-growing demand for landscape plants and trees generated by this local enthusiasm. Some businesses were short-lived, while others lasted for decades, like Germain's Seed and Plant Company, which opened in 1871 on South Main Street in Los Angeles. Founder Eugene Germain was a Swiss immigrant and influential Los Angeles resident, serving as an early vice

Into its second century, the nine-acre Japanese Garden is one of the Huntington Library, Art Collections, and Botanical Gardens' most popular spots for visitors. The elaborate gardens, introducing many plants new to Southern California, were important to the growth of nurseries and horticulture, and helped fuel the passion for ornamental plants, even among homeowners whose budgets didn't match Huntington's. Date unknown.

president of the Los Angeles Chamber of Commerce. Germain's stayed in business in large part because it kept pace with the changing fashions of gardens over the years, as plants fell in and out of favor.

Many employees of Germain's Plant and Seed Company went on to become influential nurserymen in their own right, including Theodore Payne, who had been a manager at Germain's. Payne, an Englishman, left to open his own nursery in 1902. He helped draw attention to California native plants that had been popular in England for decades, but were little noticed by Californians themselves. Payne ultimately introduced more than four hundred species of native plants and wildflowers into California landscapes, inspiring an enduring passion for wildflowers. Later in his life, he founded the Theodore Payne Foundation for Wildflowers and Native Plants, a non-profit organization that has long outlived him, carrying on his work to educate gardeners about native plants.

Payne also promoted a landscaping fad of the early twentieth century that proved less successful: a craze for eucalyptus trees sparked by newcomers from green, tree-lined Midwestern and Eastern cities who thought Los Angeles was entirely inadequate in its number of trees. The eucalyptus, first brought from Australia to California in the 1850s—probably by miners anxious to stake their claims during the California Gold Rush—appeared to be the perfect solution, growing tall and leafy in a short time, while providing wood for fence posts, furniture, and railroad ties as an added bonus. Eucalyptus became even more popular after 1904, when the federal government released projections that there were not enough hardwood trees to provide the lumber that America would need in the coming decades. Suddenly, it seemed that eucalyptus could turn a profit as well as beautify the landscape.

Planting the trees became an obsession throughout Southern California, as Angelenos made plans to cash in on this imagined bonanza by selling eucalyptus wood. By 1912, "50,000 acres of land in Southern California were devoted solely to the raising of eucalyptus. Backyard gardeners became eucalyptus merchants overnight; eucalyptus

▼ Henry Huntington's mansion was surrounded by his remarkable rose garden. Opened to the public in 1928, the home he shared with wife Arabella now houses more than 1,200 European art objects including two of its most famous paintings: Thomas Gainsborough's *Blue Boy* and Thomas Lawrence's *Pinkie*, which face each other in the main gallery. 1918.

peddlers swarmed all over the state," wrote Padilla. Theodore Payne was at the center of this movement, selling great quantities of both seeds and seedlings. The most popular eucalyptus was *Eucalyptus globulus*, known by its common name, Blue Gum. Payne and other nurserymen tried out and marketed many other species of eucalyptus as well. Like palms, eucalyptus became a ubiquitous feature of the landscape.

Another high-profile eucalyptus promoter, Abbot Kinney, is best known as the founder of "Venice of America," an ambitious development that included canals, gondolas, and amusement park attractions on the coast near Los Angeles. Less well known is Kinney's role as chair of the California Board of Forestry in the 1880s, a job he used to promote eucalyptus planting on a large scale. He also established the nation's first forestry experiment station, near Santa Monica, where he tested many varieties of eucalyptus; the station was active until the 1920s.

As it turned out, despite all the intense promotion, eucalyptus wood was not good for much besides firewood, since it warped and cracked. The trees were the bane of city maintenance crews, thanks to roots which frequently broke water pipes. Even worse, they dropped their brittle branches, occasionally injuring or even killing anyone unlucky enough to be walking below. In modern times, Blue Gum is considered invasive, outcompeting native plants, as well as creating fuel for wildfires. Though found around Los Angeles, Blue Gum is seldom planted.

Like the eucalyptus, ostrich farms and similar attractions faded in popularity, and the "fantastic farms" were mostly gone by the mid-twentieth century. Even so, they helped to build a tourist industry that was central to the Southern California economy. Gardening, as well as tourism, continued as a key ingredient of the region's growth. As the area's real estate development exploded, so too did the demand for flowers, shrubs, grass, and trees to adorn homes and businesses. Fed by this seemingly limitless growth, the nursery and landscaping industry would eventually overtake other sectors of Los Angeles agriculture. Gardens haven't lost their seductive power in Los Angeles, a place where almost anything grows, from the mundane to the exotic, from desert plants to tropical foliage, and everything in between. Whether they garden on apartment balconies or in expansive backyards, Angelenos have always created their own small paradise in the Land of Sunshine.

➤ A young boy stands among the trees in Abbot Kinney's eucalyptus grove at the Santa Monica Station. As chairman of the California Board of Forestry, Kinney was heavily involved in promoting eucalyptus. The station became part of the University of California until it was deactivated in the 1920s. c. 1900.

THE CAHUENGA SUBURBAN.

An Illustrated Monthly descriptive of

The Cahuenga and Santa Monica by the Sea.

One Dollar per Year.
Eastern Subscribers, 50 cents.

APRIL, 1896.

PRICE FIVE CENTS

A HOLLYWOOD LEMON GROVE.

8 Little Farms and the Los Angeles Agricultural Empire

Horses trotting along the city street carrying riders in cowboys hats are not what most people expect in Compton, a city better known for its gritty urban culture than its rural roots. However, one Compton neighborhood, Richland Farms, has been home to enthusiastic equestrians and backyard farmers with oversized lots. The unique community, a throwback to the era from 1910 to 1930 when suburban mini-farms helped cement Los Angeles County's status as the nation's top agricultural county, has refused to give up on its agrarian heritage.

"Compton, California, was named the eighth most dangerous city in America last year. It's the city that put gangsta rap on the map and turned Doctor Dre, Ice Cube, and Easy-E into household names," said NPR's *Weekend All Things Considered* host Guy Raz, when he visited the community in 2011. "But in one little corner of the city, you're more likely to hear roosters crowing than the bark of a 9 mm. Welcome to the other Compton: Richland Farms."

Richland Farms has received a lot of favorable attention after being a place forgotten, since it appeals to a nation intrigued by the possibilities of farming in the city. After the financial crisis of 2008, urban farming became popular as a new generation of Americans discovered that in times of economic instability, producing their own food could be especially satisfying. Attention garnered by urban agriculture initiatives, in cities such as Detroit and New York, along with excitement generated by First Lady Michelle Obama's White House vegetable garden (the first since the F.D.R. administration), helped make farming fashionable. The existence of Compton's tiny agrarian zone resonated in

The *Cahuenga Suburban* newspaper depicts farming life in Hollywood. Farmers who owned minifarms in the suburbs around Los Angeles were often referred to as "Little Landers." 1896.

▲ Richland Farms resident Derrick Jennings gets one of his horses ready to teach Compton kids all about horse bridles. Richland Farms, the last agrarian area in the City of Compton, is a ten-block neighborhood that is home to backyard farmers and equestrians. 2015.

Los Angeles and beyond. Media stories implied that if farming can thrive in troubled Compton, it can succeed anywhere.

But the story of Richland Farms is more about race than roosters, more about subdivisions than stables. Richland Farms has existed for more than a century, a place designed, built, and sold to offer a country-flavored lifestyle to white newcomers from the Midwest, conveniently close to the growing city of Los Angeles. It's a slice of a mostly forgotten, early twentieth-century Los Angeles where communities were created to sell land in the form of mini-farms. What's unique about Richland Farms is that it exists, while most such places have long since been subdivided into smaller lots, or built out into strip malls.

Neighborhoods like Richland Farms were once common throughout Los Angeles County. Especially from 1910 to 1930, new suburban developments were planned and sold with an eye toward newcomers who wanted a tiny farm to call their own, within easy reach of the nearby city. "Small farm homes," or "little farms," had plenty of support from Los Angeles boosters, including the *Los Angeles Times* and the Los Angeles Chamber of Commerce.

Richland Farms is a prototypical "little farm" community, where homes sit on oversized lots, and residents garden and raise horses, goats, and chickens, much in keeping with the original intent of the developers who named the new community with a nod to its fertile soil. "Let us show you our demonstration acre, proving the exceptional productiveness of Richland Farms soil. Vegetables, chickens, and a cow or two should net close to $600 per year on a RICHLAND FARMS acre," coaxed a 1911 advertisement in the *Los Angeles Times*. "Here is your last opportunity to buy income-producing acreage (with all city conveniences) at $650 per acre—one-third cash, balance easy payments."

The new communities springing up around the county sold a lifestyle, as much as they sold land, promoting a vision of healthful, country living, close to city conveniences. In 1910 Watts, just three miles north of Compton, Mrs. George E. Boyd enthused to a *Los Angeles Times* reporter that "living in the country within such a few miles of Los Angeles has advantages in more ways than one. Our friends who visit us from the city at our home, 'The Pines,' who praise the rich cream from our Jersey cows, who seem to be so much benefited by the ozone of the pure air, make it a pleasure to entertain them." Anyone familiar with the dense, urban Watts of the twenty-first century might be hard-pressed to imagine it as a charming country getaway.

The growth of small-farm suburbs intensified with the completion of the Los Angeles Aqueduct in 1913, transforming vast fields of dry-farmed San Fernando Valley wheat into small farms and homes with green lawns. Born in secrecy and controversy, the aqueduct resulted in massive profits for land barons, none more than members of the Los Angeles Suburban Homes Company, a syndicate that owned more than 47,000 acres in the San Fernando Valley. Its powerful five-member board included General Harrison Gray Otis, owner of the *Los Angeles Times*, as well as his business manager and son-in-law, Harry Chandler. Just as the newspaper had pushed the public to support the aqueduct, it promoted the new communities that sprang up on syndicate land, such as Van Nuys, Marian (later Reseda), and Owensmouth (Canoga Park). With the promise and arrival of water, the value of land skyrocketed.

"Under the Mighty Aqueduct a Fruitful Valley Unfolding," trumpeted a 1915 *L.A. Times* headline. "Hail, broad, fertile domain of the historic San Fernando! California salutes you in your modern garb of color and beauty! A magic transformation." The article unabashedly commended the five "far-seeing projectors"—the board of the Los Angeles Suburban Homes Company—for purchasing and "sharing" the land, rather than keeping it for their own use. It went on to credit General Otis, Chandler, and the other board members with the birth of a new, improved San Fernando Valley, where "farms, orchards, and gardens now touch edge to edge over the tract like squares of living green on a vast checkerboard."

Meanwhile, as the San Fernando Valley underwent that "magic transformation," thanks to the seemingly limitless aqueduct water, the farms at the source of the water, more than two hundred miles to the north of Los Angeles in the Owens Valley, withered and died. The project created long-lasting discord between the City of Los Angeles and residents of the Owens Valley, who on various occasions sabotaged and bombed the aqueduct that siphoned away their water.

However, the plight of Owens Valley farmers was of little concern to Los Angeles boosters. Their focus was selling Los Angeles, which suddenly had far more high-value land, thanks to the aqueduct. The small-farm movement picked up steam, as developers sold tiny farms for a premium, convincing buyers that they could make a living on a pocket-sized suburban farm. The Los Angeles Chamber of Commerce heavily promoted "little farm" development, that, according to their brochures, allowed a family to make a go of farming on two to five acres, with vegetables, fruit trees, and at least two hundred laying hens.

In reality, it was not easy to make ends meet on these tiny plots of land, so most little farm owners found other sources of income. It was in the Chamber's interest, though, to promote little farms and demonstrate their profit potential. Chamber members viewed egg production as an especially promising venture for these small-scale farmers, or "Little Landers," as they were sometimes called. To help showcase this opportunity, the Chamber courted small-farm poultry guru Charles Weeks, author of *One Acre and Independence*, a book on tiny farm self-sufficiency. Weeks had a small-farm colony in Palo Alto, a place he called Runnymede. At the Chamber's invitation, he came to the San Fernando Valley to start a new community in the west valley neighborhood known as Winnetka.

The Weeks Poultry Colony, which began in 1922, included as many as five hundred families, who paid $2500 for their one-acre homesteads. Colony families raised hens on their individual acres, then worked cooperatively to grade, pack, and market the eggs. Weeks believed that families could live well on the income from an acre of land: two or three thousand hens, raised intensively in coops, plus vegetables from a large garden.

More than just making a living, though, Weeks believed in cultivating orderly, civilized living through farming. His colony had a community center where people could get together to "meet and enjoy music and song and literature and readings and social diversions." His sales pitch for the community painted a picture of its tidy uniformity. In a 1922 *Pacific Poultryman* article, he wrote:

◄ The No Name Canyon siphon of the Los Angeles Aqueduct was dynamited in May 1927, causing three hundred feet of pipe to collapse. 1927.

This colony is a new kind of city—a city of self-supporting garden homes, all laid out on definite, symmetrical, harmonious lines. The pretty bungalows are lined up a uniform distance from the street. It is an inspiring sight to look down a street lined with orderly garden homes with an acre or two each, producing tropical growths of kale, chard, beets, giant cabbages, sudan-sorghum and alfalfa for the hens in the long uniform poultry houses.

The headquarters of the Weeks Poultry Colony in Owensmouth. Charles Weeks had established a successful poultry-raising colony in Palo Alto and was invited by the Los Angeles Chamber of Commerce to create a similar venture for Winnetka and Owensmouth in modern-day Canoga Park. 1927.

The annual Chamber of Commerce Small Farm Contest, conducted in cooperation with the *L.A. Times,* also promoted this theme of better living through suburban farming. "The one-acre farm of C.E. Drummond, 15219 Stagg Street, in West Van Nuys, is another where beauty and utility have been successfully combined in the making of a rural home. Here, again, are flowers for joy and recreation, vegetables and fruits for the table, and chickens to help swell the family purse," read one typical entry.

But there was a dark side to this vision of suburban farming bliss: it was intended for white, middle class people only. People of color, recent immigrants, and poor or uneducated people were not welcome. This narrative was clearly expressed by the *Los Angeles Times*, which wrote of the small farm home movement: "This is the real drama of this western coast. It is the story of the Anglo-Saxon farmer, settling into a semi-arid land and transforming the monotony of farming into the art of living." Charles Weeks, for his part, wanted the people in his community to be just as homogeneous as its streets. He wrote that "in the building of the Charles Weeks community only high-class people (were) selected, those living as nearly as possible on the same plane, all belonging to the Caucasian race."

In reality, most farming was far from the genteel, middle-class life that Little Landers strived to create on their suburban mini-farms. Dirty and hard, it held little comfort for the mostly immigrant labor force. Citrus workers, for example, often had no more than a pile of hay to sleep on. They handled early lead and arsenic-based pesticides without protective gear. In the winter, they breathed in the smoke from coals lit to keep fruit from freezing. Workers handled cold fruit in chilly conditions, and frostbitten fingers were common. The hip sacks used for picking, designed to keep fruit from bruising, were so heavy they often pulled men off ladders. Along with dangerous working conditions, long hours and low wages prevailed.

But despite harsh conditions, there was no end to the supply of workers, especially as thousands of Mexicans came north, recruited to California's fields due to the Mexican Revolution and its devastating aftermath of famine and chaos. With a seemingly limitless supply of low-wage labor, farms both small and large proliferated.

Many of those who came to work in the fields around Los Angeles County dreamed of one day owning their own farm. For some, especially Japanese farm workers, the dream became a reality. Japanese workers showed tremendous solidarity, organizing into associations that negotiated labor contracts advantageous to the group. In part due to this cooperation, over time they were highly successful in obtaining land and becoming farmers themselves. As their success grew, however, Japanese residents of California became the target of exclusionary legislation. One such law was California's 1913 Alien Land Act, which prohibited them from buying land or leasing it for more than three years. Some Japanese farmers found ways to work around the law, such as leasing land for short periods, or having American-born family members purchase land on their behalf.

Japanese farmers were attracted to one mode of farming in particular because it did not require owning land. Truck farming, where farmers transport their crop directly to market, was a popular temporary use of land as landowners waited to cash in on the next profitable wave of development. Japanese farmers operated on rented land throughout Los Angeles County, often employing Mexican laborers. By leasing their land for truck farms, landowners could profit, while farmers grew vegetables, berries, and other short-term crops that did not require a long-term commitment, unlike orange or walnut orchards. In this way, truck farmers made use of the bits of undeveloped land that interspersed the suburbs, boosting the productivity of local agriculture. Yet despite the key role they played in farming, the citizenry turned on Japanese farmers, just as they had the Chinese in earlier decades. According to meeting minutes, by 1919,

Chamber of Commerce members were meeting to discuss the "Japanese menace" that was "fast crowding out the small farmer."

Los Angeles Times farm writer and Chamber of Commerce staff member Ross Gast regularly demonized Japanese farmers in his articles for the *Times*, fanning the flames of hostility toward Japanese. "White men grew good vegetables cheaply before the Rising Sons arose from their rice paddies, put on their shoes, and became 'gentlemen' so they might enter this country safely," wrote Gast in a 1922 article titled "Who'll Grow the Vegetables When the Jap Moves On?" "Conditions were unfair from the start, and as the white man refused to lower his standard of living he practically left the field. Sufficient to say, the Jap was able to undersell his white competitor, and gradually force him out of business. As soon as they were out of the way, and he had control over the market, the Jap put prices up, higher than before." Blaming Japanese farmers for everything from the high price of food to the difficulty Little Landers had making a living was routine and pernicious, setting the stage for tragedy that would happen in the 1940s.

Despite this simmering resentment toward Japanese truck farmers, this mode of temporary farming was on the rise, increasingly intertwined with the sprawling Los Angeles suburbs. As the number of automobiles in Southern California grew, so too did the number of truck farms. Los Angeles County residents were enthusiastic drivers early on, with more than one hundred thousand vehicles registered by 1920. Car owners demanded more roads, resulting in more paved streets and highways that farmers could use to get their crops to market, another boost for truck farming.

The largest city in the state by 1920, Los Angeles was just beginning its phenomenal growth. The population of both the city and county would more than double during the 1920s. With population growth came soaring demand for local farm products, everything from milk and eggs, to iceberg lettuce, oranges, and asparagus. To meet the local demand, as well as ship their products around the country and the world, Los Angeles farmers found ways to produce more intensively. They added important new crops to their repertoire as well. From 1910 to 1930, their efforts would build the foundation for agriculture in California and beyond.

Dairy farming is one example of a Los Angeles industry with a lasting impact on California agriculture, thanks to Dutch dairy farmers who settled in Los Angeles County after World War I. They brought a dairy farming tradition from their space-limited homeland—corral feeding, where feed was purchased, rather than grown on site. Without the need to put precious acres into grain or pasture, dairy farms could have more cows and produce more milk. Corral feeding made sense given the high cost of suburban land, so Los Angeles dairy farmers adopted the method, which eventually spread to farms throughout California and beyond. In 1925, Los Angeles became the top milk-producing county in the U.S., as dairies expanded to meet the demand of the growing population for milk, butter, and cream. The rise of space-intensive dairies in Los Angeles County paved the way for a multi-billion-dollar California dairy industry that produced almost twenty percent of the nation's milk in 2014.

The California avocado industry, a far smaller business than dairy,

Orange groves were adjacent to an oil field in Montebello. Oil derricks could often be seen on farm land in this era. 1926.

but still a $460 million behemoth, also has its roots in L.A. While a few avocado trees had grown around Southern California beginning in the 1850s, avocados did not become an important crop until after 1910. That was thanks to the Popenoe family, owners of Altadena's West India Gardens Nursery, a dozen miles northeast of downtown L.A.

Nursery proprietor Frederick O. Popenoe and his sons collected avocado and other trees, often venturing south of the border to Mexico and Central America to find new and exotic species. After a devastating freeze in 1913 killed most of their stock at the nursery, they made a key discovery. One particular avocado, collected in Mexico, had survived the freeze unscathed while

An avocado field covered what eventually became La Habra Heights. Date unknown.

others had died. They named this tree *fuerte*, Spanish for "strong," to capture its hardy nature. With some marketing, it was soon a popular and lucrative crop among local orchardists. The Fuerte avocado originated in Atlixco, Mexico. But it was "discovered," named, and popularized in Los Angeles County, becoming the standard upon which the California avocado industry was built.

The next great avocado discovery was another lucky Los Angeles County accident. The Hass avocado (sometimes marketed as the Haas), later eclipsed the Fuerte as the California avocado of choice. It was a random unknown variety, planted in 1926 by postman Rudolph Hass in his La Habra Heights orchard. He thought its rough fruit were unappealing, and planned to cut the tree down, but his children convinced him that its fruit were actually superior. He subsequently took out a patent, naming the tree after himself. By the time the millennium had turned, around ninety-five percent of California avocados were of the Hass variety, all originating from Hass's original tree that lived seventy-six years, until it succumbed to disease in 2002.

Taking a page from citrus and walnut farmers, avocado growers formed their own cooperative marketing association, the California Avocado Exchange, which became Calavo Growers in 1927. Los Angeles-based Calavo designed marketing campaigns to convince consumers to eat more California avocados, that they touted as "the aristocrat of salad fruit." Some Calavo recipes have faded into history, the Calavo Cocktail, for instance, that involved cubed avocado and celery served with a mayonnaise and ketchup dressing, while others, especially guacamole (a traditional Mexican dish Americanized and promoted by Calavo as "avocado dip"), became international standard fare.

As marketing associations like Calavo, Sunkist, and the California Walnut Growers Association continued to thrive during this era, so too did wholesale markets. Farmers drove the development of new marketing centers such as the Los Angeles Flower District, a busy spot for modern florists and flower-lovers seeking early morning bargains on blooms. The district dates back to 1913 when Japanese flower farmers organized the Southern California Flower Market. Then in 1924, European immigrants opened the Los Angeles Flower Market across the street. The district is promoted as the largest wholesale flower market in the nation, though the flowers are no longer grown in Los Angeles.

Livestock sales, too, had a massive boost during this same era. The Los Angeles Union Stockyards, established in 1922, made Los Angeles a booming center for livestock sales that rivaled Chicago and Omaha.

The stockyards reinvigorated cattle ranching in Southern California, which had declined after the *ranchos* disappeared. It also spawned a surrounding district of feedlots where cattle were fattened before sale. While the stockyards and feedlots are gone, slaughterhouses and meat-packing businesses continue to operate in the same area, including a large Farmer John facility. Recognized by its giant mural of happy pigs on a happy country farm, the Farmer John plant is a frequent target for animal rights demonstrations.

As wholesale markets and the massive stockyards grew up near railroad lines, Los Angeles farmers could more efficiently get their products to consumers around the nation. International markets opened up as well when the harbor at San Pedro was dredged and widened in 1912. With more and larger ships coming into the harbor, farmers increasingly looked beyond the local market to national and international markets, even more so with the opening of the Panama Canal just two years later.

The trend toward scientific farming continued as well. In 1917, Los Angeles County partnered with the University of California in Berkeley to establish U.C. Cooperative Extension, an organization that continues to work with local farmers to encourage science-based farming practices. U.C. also established a new campus in Los Angeles that would play an important role in farm education and research. The Southern Branch of the University of California, later renamed UCLA, first opened in 1919 on Vermont Avenue near downtown Los Angeles. It surpassed capacity by the early 1920s, prompting a search for a new location.

After considering numerous sites, U.C.'s board of regents settled on a 380-acre chunk of land with deep roots as farmland. Part of the Rancho San Jose de Buenos Ayres, a Mexican-era land grant, the site was once owned by Benjamin Davis "Don Benito" Wilson, an early citrus grower, vintner, and the second mayor of Los Angeles. John Wolfskill, a member of the pioneering citrus-farming family, purchased it for ten dollars an acre in 1884, raising cattle, beans, and barley until his death in 1913.

Despite the land's rich agricultural history, U.C. did not intend to have any farming on its new campus, preferring to keep its agriculture program restricted to its Berkeley campus. But pressure from local farmers prevailed, and a ten-acre demonstration orchard was planted in May 1929, before the September opening of the campus. The UCLA orchard evolved into a collection used for teaching and research on subtropical fruit, including numerous species of avocado, citrus, persimmon, almond, fig, and cherimoya.

Local farmers, however, were not yet satisfied, since they wanted a full agriculture program on the

➤ The rapid growth of the Los Angeles Stockyards made them important rivals of Chicago and Omaha. 1951.

campus, beyond what a ten-acre orchard could offer. When farmers flexed their political muscles, blocking funding in the state legislation intended to fund construction at the new campus, the U.C. regents capitulated. By the early 1930s, UCLA had a full-blown agriculture program, including courses and research on subtropical horticulture, entomology, irrigation, soils, and plant pathology.

▲ Calavo Growers of California warehouse, located at 4803 Everett Street in downtown Los Angeles, was part of modern-day Vernon. At the time, this was the grower-owned cooperative's main packing plant in California. c. 1930.

➤ The California Flower Market, started by fifty-four *issei*, or first-generation Japanese, organized into the Southern California Floral Industry Association (eventually known as the Southern California Flower Market). In January of 1913, the Southern California Flower Market opened at 421 Wall Street. 1933.

The Los Angeles Chamber of Commerce, too, continued to support and promote agriculture and farming, often through pamphlets like one titled *Los Angeles: the Center of an Agricultural Empire*, published in 1929, which touted the preeminence of Southland farming. Whether in the form of big cattle feedlots, suburban dairies, or tiny one-acre farms, agriculture was key to the economy.

In many ways, though, farming was window dressing for the real business of Los Angeles: real estate. Farming fed development, and vice versa. "Bumper Crop Reports Speed Realty Sales," announced a 1929 *Los Angeles Times* article, which reported that buyers were snapping up lots in a new development called Walnut Acre Farms after the California Walnut Growers Association reported a record-breaking crop. Once they had their small walnut farms, these new growers would add to the local farm abundance. And

増築落成後ノ南加花市場
午前六時開場後乃部ノ模様（南側）
May 20-1933

likely, their purchase was as much about buying a lifestyle as it was about starting a business.

In Compton's Richland Farms, it's more about lifestyle than commerce, since long-time residents raise horses and livestock mostly as a hobby. While the area was segregated for decades, African American families began to move into the formerly all-white neighborhood in the 1960s. Their children grew up riding horses and participating in 4-H Clubs, creating a rural life in the city that for many, harkened back to their roots in other parts of the country. "I'm a country boy. I come from a farm. That's what made this place special for me. I felt like I was back home," one longtime resident, formerly from Texas, told a *Los Angeles Times* reporter. And since then, Latino residents have joined

Portuguese Bend on the Palos Verdes Peninsula was known for its garbanzo beans and barley, as well as the hundreds of acres of strawberries grown by Japanese American farmers who leased parcels of land from the Bixby family. 1920.

Workers pose in front of the California Walnut Growers Association packing plant. The Association had two plants in the Los Angeles area; this was likely the one in downtown Los Angeles. c. 1920.

the community, many of whom also come from rural areas. Horseback riding is a passion shared by many residents regardless of their background. Richland Farms is home to a prize-winning youth equestrian program, the Compton Junior Posse, that has trained a new generation of cowboys and cowgirls.

But the neighborhood has faced challenges related to its rural qualities, including underground backyard milk sales and cockfighting, leading city officials to put restrictions in place, such as reducing the numbers of animals allowed on each lot. In addition to concern about these rules, Richland Farms residents have worried over the city's tendency to allow construction of additional homes on each property. Some residents believe that their unique community is threatened by city officials who are making it easier to build and more difficult to keep animals. "They are sabotaging the community so that developers can eventually come in and take over," according to one resident interviewed by the *Los Angeles Times*. If that's the case, it won't be the first time that farming has lost out to development. Once the center of an agricultural empire, Los Angeles has always also been a paradise for the developers, builders, landlords, and realtors who stand to profit from an ever-denser urban landscape.

NEXT TIME TRY THE TRAIN
RELAX
Southern Pacific

9 Farming in Depression-Era Los Angeles

El Monte, in eastern Los Angeles County, has billed itself as "the End of the Santa Fe Trail" to tout its pioneer heritage. Its place in California farm worker history, however, is mostly forgotten, submerged under pavement and the passage of time. It was once ground zero for a worker uprising that spread throughout the county to become one of the largest California agricultural strikes of 1933, a year of widespread farm labor unrest. As the Great Depression unfolded in Los Angeles, farm laborers throughout the county organized to fight for better wages, decades before Cesar Chavez made farm worker rights a national topic of conversation. Farmers, too, struggled, trying to hang on to their farms. As the economic crisis deepened, the popular mini-farms of previous decades became less about cultivating a country lifestyle in the suburbs, and more about survival.

Events in El Monte foreshadowed a decade of conflict in the fields around Los Angeles. The El Monte Berry Strike, also known as the Los Angeles County Strike of 1933, was centered in El Monte's Hicks Camp, a barrio where farm worker families survived by following the crops. At tiny Rio Vista Park, the site of a memorial to Hicks Camp and El Monte's forgotten history, the past speaks through the words of Richard "Huicho" Perez, etched on a plaque there.

"I started working in the fields when I was seven years old. . .The season started in June with our route beginning in Salinas/Castroville to bunch carrots; then San Jose to pick prunes; then Livingston to pick grapes; then Corcoran to pick cotton; then Moorpark/Simi to pick walnuts, and finally back to El Monte...

Towards Los Angeles by famed Depression-era photographer Dorothea Lange, captures the rootlessness of the transient Americans who flocked to Los Angeles searching for work. 1937.

Farm workers pick raspberries in Rosemead, near El Monte, where berry pickers would be at the center of a huge agricultural strike by 1933. 1927.

Our working days were sunrise to sundown. We would take a half hour break in the middle of the day to eat. Every day was a working day. Weekends and holidays were unheard of," said Perez about life at Hicks Camp.

The grindingly hard life of families like Perez's helped to make El Monte "the larder and granary" of Los Angeles, known for its productive farms and a cornucopia of crops. "If this valley were suddenly converted into a Nevada desert, many infants would be stinted in their daily milk, many hot waffles would go butterless, many vegetable dishes would accumulate dust on pantry shelves, the potato kings would have fewer spuds to corner, and the price of walnuts would take a big jump in every eastern city," according to a 1911 *Los Angeles Times* article.

The workers who made this abundance possible were mostly Mexican and lived in barrios like Hicks Camp. With small houses cobbled together from salvaged lumber and a lack of streets or city services of any kind, Hicks Camp typified the communities where farm workers lived. Scarcity was the norm, but as the Depression took hold, farmers cut wages in response to declining prices, pushing farm workers further into poverty.

In the spring of 1933, worker discontent in El Monte came to a head during the raspberry and blackberry harvest. With just a brief season, berry picking was short-term employment, when families—men, women, and children—all worked in the fields to maximize their earnings. Farmers paid pickers by the crate, which meant about twenty cents an hour for a skilled worker, or as little as nine cents for children and others who worked slowly.

When organizers from the communist-affiliated Cannery and Agricultural Workers Industrial Union (CAWIU) arrived in El Monte in May 1933, they found fertile ground for organizing. Hicks Camp became their headquarters. "500 Raspberry Pickers Meet; Plan Fight," wrote the *Western Worker*, a communist party newspaper, in a short article dated May 29, 1933. A week later, the *Worker* wrote "fifteen hundred pickers from San Gabriel Valley went out on strike June 1 against starvation wages and miserable working conditions."

Within days, as many as five thousand workers went on strike, as the walk-out spread to fields in Venice, Santa Monica, and other parts of the

A farm-working family harvests tomatoes in the San Fernando Valley. It was common for children to work alongside their parents in the fields. c. 1930.

Workers size and pack in a Los Angeles County citrus-processing house. 1939.

county. A strike of this magnitude was unusual, especially near Los Angeles, a place some called "scab city," where the business community kept a tight lid on union organizing. In a decade when farm workers increasingly demanded higher wages and fair treatment, the El Monte Berry Strike was the first major farm strike in Los Angeles County, but far from the last.

Although the strike was largely peaceful, as it progressed, the anti-union *Los Angeles Times* highlighted the negative aspects of strike activity, as seen in an article titled "Berry Strike Gets Violent: Radical Group Ordered Out of El Monte Area." According to the article, "a group of strikers and agitators stopped a truck at Marshall and Walnut streets, Rosemead, and pulled Pedro Zuniga off. One of the gang beat him with his fists, and deputy sheriffs from the Temple substation arrested the asserted assailant, who gave his name as Danny S. Cardiel of Hicks Camp." In a remarkably speedy trial by modern standards, a jury found Cardiel guilty of assault and battery just three weeks later, when he admitted that he hit the other man because Zuniga had broken the picket line.

As the strike went on, farmers fought back by turning their fields into "pick your own" operations, using radio and newspapers to invite families throughout the county to come pick berries for a penny a box. Under the watchful eye of sheriff's deputies, "hundreds of men, women and children carrying baskets, buckets and various other containers invaded the berry fields in El Monte and Arcadia today to pick the crop remaining on the vines, a result of the strike of Mexican berry pickers for increased pay which has been in effect for some weeks," according to a June 29, 1933 article in the *Los Angeles Times*.

Despite many challenges, the striking workers hung on, encouraged by local leaders like Zenaida (Sadie) Castro, a resident of Hicks Camp who brought food to those on the picket line, as well as coverage in the *Western Worker*, which published a poem in honor of the strikers: "Hail you workers in El Monte, in your fight you're not alone. We are one with you in struggle and your deeds are not unknown."

The strike highlighted the racial and power disparity between the various factions, each having a vested interest in its outcome. In what had become a common scenario in Los Angeles County, the workers were mostly Mexican, and the farmers were primarily Japanese. Less visibly, the land itself was owned by white landowners who rented to Japanese farmers, often quietly extending leases past the three years allowed by law under the Alien Land Acts. Since these arrangements were profitable for landowners, the Los Angeles Chamber of Commerce helped to mediate the strike, because it did not want to call public attention to the over-extended lease status of many Japanese farmers. By July, the

When citrus groves were sprayed to protect against pests, unprotected farm workers were routinely exposed to pesticides. 1937.

strikers and farmers reached a settlement for a small wage increase and pickers went back to work. However, with much of the harvest over, there was little real benefit for the workers.

The fact that Mexican farm workers were willing to risk going on strike was a testament to both their courage and their desperation, given events of the early 1930s. With anti-immigrant sentiments at a fever pitch, deportation was a very real threat. Just as Chinese and Japanese immigrants had been vilified in prior decades, as the Depression worsened, Mexicans and Mexican Americans became the focus of hostility, blamed for taking jobs away from "real" Americans.

Just a few years earlier, California farm labor contractors had aggressively recruited Mexicans for farm jobs in Los Angeles and around the state. Los Angeles was a center for farm labor recruitment, with numerous offices that signed up workers to pick cotton and other crops in the Central and Imperial Valleys. Now, the very community that had encouraged Mexicans to make a new life in California turned on them, preparing to return them, whether voluntarily or by force, to a homeland that many no longer knew.

Beginning in 1931, Los Angeles County officials decided that in order to save money on the county's rapidly expanding public assistance rolls, they would transport Mexicans living in Los Angeles to Mexico, a campaign they termed "repatriation." Such efforts were common around the U.S. during the 1930s, but "Los Angeles County became the hotbed of the repatriation movement," according to Francisco Balderrama and Raymond Rodríguez in their book *Decade of Betrayal: Mexican Repatriation in the 1930s*. The county paid the Southern Pacific $14.70 per person to take trainloads of its residents to Mexico, sometimes after rounding them up in public places such as parks. Officials also pressured residents to leave for Mexico voluntarily. Social workers, for example, encouraged clients with Spanish surnames to

▼ A Mexican worker picks oranges. Date unknown.

This Federal Writers' Project photo shows workers packing avocados at the Calavo Growers plant in downtown Los Angeles. Date unknown.

participate in repatriation, suggesting that they would lose their public relief benefits if they stayed, even when it was not true. Often, Mexican-born parents felt that they had little choice but to accept the offer of a train ride to Mexico, taking with them children who were U.S. citizens. The campaign divided countless families over the course of the decade, separating children from parents, husbands from wives, as some family members stayed and some departed, whether voluntarily or by force. Once in Mexico, the "repatriates" had little opportunity in a country still struggling to recover from a devastating revolution.

The first major raid in L.A. County's decade-long deportation effort was held in El Monte's agricultural areas, just two years before the Berry Strike. Despite the threat of deportation, Mexican workers, along with Filipino, Japanese, and other low-wage farm laborers, continued to organize after the 1933 El

Burlap sacks are filled with thousands of walnuts in this packing plant in La Puente. Date unknown.

Monte strike. Around the state, the demands of farm workers soon provoked a coordinated response from a powerful group called Associated Farmers, whose backers included prominent members of the Los Angeles business community, such as Sunkist President Charles Collins Teague, Harry Chandler of the *Los Angeles Times*, representatives from railroad companies and utilities, and various members of the Los Angeles Chamber of Commerce. Associated Farmers worked with local sheriffs' departments, encouraging vigilante groups to quell farm strikes, often with violence, while lobbying county boards of supervisors and the state legislature against pro-labor measures.

Despite heavy-handed efforts to subdue unions, Los Angeles County agricultural strikes increased in frequency and violence as the decade progressed. In one instance, a 1936 Los Angeles vegetable workers' strike, sometimes known as the Venice Celery Strike, also had a mostly Mexican workforce facing off against Japanese farmers. Many of the Japanese farmers wanted to agree to the workers' demands, but were deterred from doing so by organized farm groups, who urged the Japanese growers association to "fight it out," according to Carey McWilliams in his classic book on California farm labor, *Factories in the Field*. Organizers set the strike to start during the harvest of celery, an important crop with several thousand acres grown in a swath of coastal land between Culver City and Venice.

Twenty-eight year old Lillian Monroe of the Agricultural Industrial Workers Union (AIWU), described as a "radical labor agitator" in the *L.A. Times,* led the striking workers. Over the course of the strike, *Times* reporters described various skirmishes in the celery fields, with headlines such as "Tear Gas Bombs Rout 300 Strike Rioters in Venice." The *Times* regularly lambasted Monroe, who would go on to organize citrus and cotton worker strikes, as a communist rabble-rouser. Many strikers were injured, some seriously, mostly from being tear-gassed and beaten. During the month-long strike, 1,500 police, sheriffs, and armed guards arrested as many as a third of the two thousand striking workers.

Strikes in the dairy industry were especially common during the second half of the decade. In one such 1937 strike, dairy workers were picketing for a $120 minimum monthly salary, an eight-hour day, and a six-day workweek. The *L.A. Times* reported: "Deputy sheriffs stood guard over a stack of 60,000 bales of hay at the Vitamin D Dairy in Norwalk last night as threats to burn the pile marked the latest development in the dairy's labor troubles." The *Times* also reported instances in which striking workers forced milk trucks off the highway on their way to market.

The intensity and violence of farm worker strikes increased in part due to the flood of migrants from Oklahoma and other Dust Bowl states who came to California looking for work, overwhelming the farm labor market and further driving down wages. Dust Bowl migrants soon became the new scapegoats blamed for the ongoing economic woes of the agricultural sector throughout the state. Ironically, decades of successful Los Angeles Chamber of Commerce campaigns promoting California as the land of sunshine, with

The Original Farmers Market opened in 1934 as a way for local farmers to sell their fresh produce during the Great Depression. Despite the fact that it no longer sells exclusively locally grown produce, The Farmers' Market remains a popular tourist attraction in Los Angeles. Date unknown.

fruit on every tree, were a huge part of what drew the desperate Dust Bowl migrants west. John Steinbeck's *The Grapes of Wrath* vividly portrayed the struggle and desperation of these unwanted migrants, often known derisively as "Okies." But like Steinbeck's fictional Joad family, Dust Bowl refugees soon learned that California was no paradise, with both fruit and work mostly unattainable as gun-toting guards kept a close eye on orchards, while candidates for farm jobs overwhelmed demand.

Despite the oversupply of farm labor around the state, the powerful farm lobby, led by Associated Farmers, convinced Los Angeles County relief agencies that families on public assistance ought to be working the harvest, instead of "on the dole." Officials ordered thousands of Los Angeles-area families to report to pick crops, traveling at their own expense, sometimes up to hundreds of miles away in other counties. Some returned to Los Angeles to protest horrendous conditions in the fields, rioting at Los Angeles-area relief offices. Despite such protests, farm organizations around the state continued to successfully demand that Los Angeles County officials send thousands of workers to harvest crops.

Given the challenges and stigma associated with accepting help from public or private sources, some Depression-era groups organized around alternatives to charity. One example was the self-help cooperative movement, with its motto "Self-help beats charity: Charity is for abnormal people in normal times; we are normal citizens in abnormal times." Starting in Los Angeles County and spreading around the U.S., self-help cooperatives were based on bartering labor for goods, for example, harvesting crops for a share of the harvest. The first cooperative was organized in Compton in 1932. Within a year, forty-five had been established in L.A. County.

The self-help movement became active in other parts of the U.S., but was strongest in Southern

In 1934, there were 14,000 members in Los Angeles County who harvested crops, milked cows, gardened, and did other work in exchange for food. Date unknown.

NO
WORK
EAT
TRADE UNION LEAGUE
ANNUAL
PICNIC
SUNDAY JULY
AT LANCASTER LAKE
SUNLAND CALIF
HAIR CUTS
FREE
DONATION

California, especially Los Angeles County, with 14,000 members there at the peak of the movement in 1934. Members usually worked sixteen hours per week in support of their cooperative and its projects, and in exchange, received food, clothes, and other goods. Cooperatives did everything from baking, to canning, to fishing, but farming was one of the most common activities. At the cooperative in Santa Monica, members worked at local dairies in exchange for milk and tended the research orchard at UCLA for payment in fruit. The Huntington Park Cooperative had vegetable gardens on vacant land, harvesting more than 27,000 pounds of fresh produce in just one quarter of 1936.

Survival on the land became an even larger focus of the Small Farm Home movement that had been so popular throughout the 1910s and '20s. It gained momentum during the Depression, becoming entwined with a New Deal program called Subsistence Homestead that aimed to create communities where low wage city workers would become self-sufficient on their own tiny farms on the outskirts of cities. Los Angeles boosters felt a keen sense of ownership in this new program. "It is no secret that Uncle Sam's representatives made a very careful study of our small-farm homes before they launched their national program," said a 1935 *L.A. Times* article. While the Los Angeles business community had no great love for President Roosevelt and his New Deal, they saw the Subsistence Homestead Program as a perfect match for Los Angeles, since it focused on a favorite theme: self-sufficient living on a small plot of land.

While organizations and individuals throughout the nation lobbied Washington to be chosen for the Subsistence Homestead Program, Ross Gast, the bombastic *Los Angeles Times* writer and Chamber of Commerce Small Farm Home promoter, was one of the most successful advocates. He spearheaded the first successful bid to have a Subsistence Homestead Program west of Texas, scoring two such projects for Los Angeles County, one for the San Fernando Valley community of Reseda, and one for El Monte. As part of his efforts, he popularized the term "rurban" to describe the small-farm lifestyle on the edge of the city, offering the best of both country and city life.

With an anticipated one hundred homesteads, each on three quarters of an acre, the larger El Monte project sparked widespread excitement, generating thousands of applications. Participants were selected from an all-white pool of low-wage workers. Mr. and Mrs. M.J. Evans and their five-year-old son Jimmy were the first family to move into their new homestead. Mr. Evans, employed by an auto supply company in Los Angeles, was similar to many of the new homesteaders. These "rurban" farmers worked in industry

◄ Depression-era photographer Dorothea Lange photographed a family working on their El Monte small farm homestead, part of the national Subsistence Homestead Program. 1936.

Women in a Los Angeles area cooperative canning peaches. During the Depression, because the price a farmer could get for his crop might not cover the cost of harvesting, co-ops often did the harvesting in exchange for a share of the harvest. Date unknown.

and commerce, with little experience in farming. They were often more interested in the cheap mortgages the program offered than in growing crops, but were required to farm their new land regardless.

For the families who found refuge in the homesteads, or held on to their small farm homes, the food they grew and the roof over their heads sheltered them from the worst deprivations of the Depression. For most farm workers, though, conditions continued to worsen. Hunger, even starvation, was common as workers and their families struggled in shantytowns that sprang up in agricultural areas around California. People created makeshift housing from cardboard, gunny sacks, and other items, and lacked even minimal shelter. "People are seeking shelter and subsistence in the fields and woods like wild animals," said one observer. For those lucky enough to find work in the fields, wages might be as little as fifty cents a day. By 1937, state officials were investigating conditions in the fields of California. Reports described "unimaginable filth," and housing "unfit for human habitation," along with epidemic influenza, diarrhea, and typhoid fever.

Along with the *Grapes of Wrath*, the 1939 publication of *Factories in the Field*, Carey McWilliams's exposé on California farm labor, helped to bring the plight of migrant farm workers to the national stage. McWilliams, a Pasadena lawyer who took up the cause of California farm workers, was persona non grata among the state's powerful farm groups. "Of all the pests which the crops of California are infested with, Mr. McWilliams is Agricultural Pest Number One," wrote the Associated Farmers. As the 1930s ended, however, Governor Culbert Olson appointed McWilliams to oversee housing and immigration for the state. The tide began to turn against Associated Farmers as the governor sided with McWilliams in supporting basic farm worker rights. The economic situation rapidly improved as new jobs in the industrial sector began to open up. With new, better paying jobs, people could again afford to buy fruits, vegetables, and other farm products, helping farmers to regain their footing after the hard years of the Depression.

At the close of the turbulent decade, farming in Los Angeles seemed relatively unscathed. According to a 1940 Los Angeles Chamber of Commerce promotional brochure, "Los Angeles County continues to hold first place as the richest agricultural county in the entire nation. It has 13,549 separate ranches and farms (their average size is 45.7 acres) which each year produce from $75,000,000 to $90,000,000 in farm products."

In twenty-first century El Monte, things have changed on the surface. The farms, of course, are long gone. With its quiet residential areas and businesses—a Home Depot here, a Starbucks there—El Monte seems to be a peaceful, prosperous suburb. But decades after the El Monte Berry Strike, an undercurrent of need has remained. In 2013, El Monte had a higher poverty and unemployment rate than the state average, with many of the employed working in low-wage jobs such as those in the fast food industry. In many ways, fast food jobs had become the new edition of farm labor in urban communities, with wages so low that employees often had to work multiple jobs and go on public assistance. Pushed to their limit, fast food workers organized protests in El Monte and two hundred cities nationwide in December of 2013.

"On strike to lift my family up," read the words on a handmade sign at one such protest outside of a South Los Angeles McDonald's, where picketers chanted "keep your burgers, keep your fries, make our wages super-sized." According to a 2013 U.C. Berkeley study, more than half of the nation's fast food workers relied on public assistance programs like SNAP (food stamps), meaning that taxpayers shouldered a significant cost for the low wages paid by fast food restaurants.

Meanwhile, on California's farms, the issues have remained strikingly similar to those of the 1930s, including low wages, harsh working conditions, and fear of deportation among the many undocumented, mostly Mexican workers. Despite their role in the land of plenty, many farm worker families have struggled to put food on their own tables, unable to afford the very produce they harvest. Whether the job is picking berries or flipping burgers, the struggle for a living wage has continued among food industry workers from the streets of Los Angeles to the fields of Central California.

10 Farming in World War II and Postwar Los Angeles

As the Great Depression faded, Los Angeles entered the era when it would lose the majority of its farms. Farmland disappeared under highways, homes, parking lots, and pavement, as the suburbs spread out, after the end of World War II. During the war years, food production was a national priority, which sustained local farming, as well as sparked a Victory garden movement at schools, in backyards, and on vacant lots throughout Los Angeles. But as the war ended, the focus on agriculture faded. Southern California entered a massive post-war boom when its citrus groves and farm fields fell to bulldozers in ever-increasing numbers. Between 1940 and 1960, Los Angeles completed its transition from cow-county to sprawling metropolis, as farmers were either bought out or forced out.

For the county's many Japanese farmers, the losses were devastating and abrupt. After the December 7, 1941, attack on Pearl Harbor, Japanese farmers soon became a renewed focus of hostility. "Japanese Menace," screamed a March 1942 *Los Angeles Times* article titled "Network of Japanese Farms Covers Vital Southland Defense Areas." A map of the county's many Japanese-owned farms was included, "depicting how Japanese landholdings are spaced throughout the Los Angeles County area in a manner to permit disastrous assaults on every military objective."

With wartime paranoia at its height, many believed that Japanese farmers were spies, ready to assist the enemy from their strawberry and vegetable fields. By the end of March, Japanese families were being evacuated, first to assembly centers, then to internment camps, a result of Executive Order 9066, authorizing the forcible removal of Japanese Americans from the West Coast.

Chiyeko Suzumoto harvests red cabbage at a truck farm in Gardena.1951.

Japanese workers on a vegetable farm near the Lockheed plant in Burbank. After Pearl Harbor, Japanese operated farms like these were transferred to non-Japanese farmers, due to Executive Order 9066, which allowed government officials to evacuate and intern Japanese and Japanese Americans on the West Coast. Japanese families often lost their farms, and many were unable to return to farming after the war, since they had lost not only their land, but equipment and other investments. 1942.

"My father was a truck farmer," said Hiroshi Mayada in an interview years later about his family's experience losing their farm due to internment. His father farmed in San Pedro, growing tomatoes, celery, strawberries, and flowers that the family sold at a successful roadside stand. "Things got going real well. Then Pearl Harbor happened. My father saw the war clouds, and he was preparing to move on his own, inland, like some Japanese did in those days. They went to Colorado, or Utah, or Idaho, but the evacuation orders came. He had sold off a lot of the farm equipment we had: trucks, and a John Deere tractor, and a Caterpillar tractor, and two horses and a mule, all kinds of things like that. Then we went into Jerome Relocation Center in Arkansas and, at that time, I was just barely sixteen and consequently, I finished high school in camp." Like many of those interned, Mayada, born in Compton, was an American citizen.

Given wartime pressure to produce more food, Los Angeles County created a special agency to find new farmers for thirty thousand acres of land that had been cultivated by Japanese growers like Mayada's father. Officials coordinating this effort reported, "We have found that a surprising number of Japanese have been renting tracts for fifteen to twenty years on a month-to-month basis without ever having signed formal leases." Of course, this was really not so surprising, since state law restricted Japanese ownership

Frank M. Kramer, administrative assistant to the State Agricultural Commissioner, poses with Japanese-owned farm equipment in storage in Culver City, enough machinery to farm three thousand acres. He appeared before the Senate Committee on Military Affairs demanding it be released to county farmers who needed it. As a result of his efforts, representatives went to Manzanar to buy the equipment from the Japanese farmers. 1943.

The City Market at Ninth and Wall Streets, just south of Little Tokyo, was photographed just five months prior to the Japanese attack on Pearl Harbor. July, 1941.

UNITED GROWERS
Y. KUBOTA
Y. KUBOTA
IRON BRIDGE RANCH
JUMBO
MELONS
VEGETABLES

of land and limited leases to less than three years. Without formal leases or other legal agreements, government officials could easily reassign land that Japanese families had farmed for years, or even decades.

In the spring of 1942, farms were transferred from Japanese growers to the farmers taking over their land in mid-crop cycle. Japanese farmers lost the seeds, fertilizer, labor, and other resources that had gone into production. One such transfer was described in the *Times* on April 8, 1942: "A typical transaction is that involving forty acres across the street from a major Burbank war plant. The property, planted to root vegetables, cabbage and strawberries, was taken over by L.A. Gilbert and his son, Harold B. Gilbert. A workhorse and some farm machinery were included in the deal. Three Japanese families formerly lived on the site within a stone's throw of the factory. They moved away last week and since then the Gilberts have employed Mexican hands to harvest the flourishing crop of carrots and radishes."

Mexican nationals came to Los Angeles by train to fill wartime labor shortages on farms. The Bracero Program began in 1942 and continued until 1964, with many Braceros working on citrus farms around Los Angeles County. 1942.

Beyond reassigning Japanese farmland, a new challenge to wartime farming soon presented itself: there simply weren't enough farm workers. As new opportunities developed in the expanding defense industry, fewer workers were available for low-paid agricultural work. By fall 1942, many of the growers who had taken over Japanese farms were complaining that there was no one left to pick the crops. Local papers were full of stories of office workers and high school students picking crops, as they stepped in to fill the gap. "Typewriting is a cinch compared to picking cucumbers. So say the Hollywood secretaries who each weekend help harvest Victory crops in the San Fernando Valley," said a *Times* article in summer 1943, which encouraged "Victory harvesters" to sign up at their local Farm Labor office to help pick tomatoes, grapes, walnuts, and "the general run of truck crops."

While students and secretaries picking crops for victory made good press, the labor gap was far too wide to be filled by part-time workers. With a farm worker shortage all over the country, federal officials created a new program to import a fresh supply of cheap farm labor. While Mexicans had been blamed for the lack of jobs during the Depression, with thousands "repatriated" to Mexico, the new Bracero Program provided a pathway for Mexican nationals to work in the U.S. from 1942 until 1964. During that time, more than four million Mexicans would join the program to work on U.S. farms, coming from agricultural areas of Mexico, leaving their homes for the promise of opportunity. Over the years, thousands of Braceros worked on Los Angeles County farms, especially in its citrus groves.

While farmers struggled with labor challenges in maintaining the food supply, Angelenos renewed their passion for gardening in response to the national push to plant Victory gardens that would help fill the gap. "Every pound of food raised and used at home saves a pound that would otherwise be purchased in stores or markets and thus diverted… from shipments to our allies, or military and naval supplies," pronounced government officials in January 1942. Around Los Angeles, Victory gardens sprouted up in backyards, vacant lots, schoolyards, college campuses, and anywhere else that a plot of vegetables—large or small—could be planted.

The *Los Angeles Times* jumped on the Victory garden bandwagon in a big way, creating its own Victory Garden Club that had 25,000 members by October of 1943. The *Times* put in a demonstration Victory garden on Wilshire Boulevard, as well as a one-acre Victory Garden Center on Adams Boulevard in downtown, that featured a garden library, a lecture room for Victory garden classes, and a supply of gardening tools that could be checked out by club members.

Cities set goals for numbers of gardens their residents would grow. The small city of Alhambra, east of downtown Los Angeles, for example, targeted six thousand gardens in 1944, "one for every seven residents." The City of Los Angeles found many ways to promote gardens, such as selecting a local "Victory Garden Girl" as a spokesperson for gardening, and proclaiming an annual Victory Garden Week. High school

Farmer Alfred Rodriquez plows a hillside field in Palos Verdes using mule power. 1951.

Advertising the Los Angeles Times Model Victory Garden, located on Wilshire Boulevard in the Wilshire District, was part of the *Times*'s effort to encourage wartime Victory gardens. Date unknown.

Standard Oil of California had its offices in downtown L.A. on W. Olympic Boulevard, and produced a series of free Victory garden guides encouraging the public to grow "Food for Freedom." Like other oil, chemical, and fertilizer companies that produced similar guides, they encouraged Victory gardeners to use their products . Date unknown.

A yearbook page commemorates the success of the 1942 Victory gardeners at Manual Arts High School. 1942.

student Marie Denham, the Victory Garden Girl of 1945, was presented at the mayor's reception room at City Hall, "dressed in denim, with a rake, hoe, and vegetables," according to the *Los Angeles Times.*

Students at every level caught the Victory garden bug. At UCLA, students planted a garden north of the Women's Gym, as the Kerckhoff Hall cafeteria pledged to buy every tomato, carrot, and head of cauliflower grown there. Local elementary and high schools also promoted Victory gardens, both at school and at students' homes. L.A. Unified School District provided free advice to local gardeners at all four hundred city schools, offering garden clinics to assist in their "garden-in-every-backyard program."

Even Hollywood songwriters helped promote Victory gardens, penning songs that were popular throughout the nation. One such song, 1944's *We've Gotta Dig for Victory*, included lyrics such as "Don't hang around that hotel lobby / Go and plant some fresh kohlrabi / Make your garden work your hobby /

At the Homes

▲ A couple poses in their Dalton Avenue Victory garden for the *Los Angeles Herald-Examiner*. They raised twenty-six vegetable varieties for $8.60 and "grew enough fresh greens to keep their own household and half of the neighbors supplied." 1942.

We've gotta dig for Victory!"

Victory gardens were successful in wartime Los Angeles in part because they were a perfect use for vacant lots sitting empty while investors anticipated a post-war development boom. The *Times* reported: "Throughout the Los Angeles region hundreds of vacant lots are blossoming and will bloom into Victory gardens, not only through cultivation of many of them by owners but also because sites are being loaned for such purpose. Hundreds of investors buying lots for building after the war are meantime turning them into Victory gardens."

According to the school district, "Richard Russell and Bette Tubbs learn how to raise rabbits in the animal husbandry class at school on April 28, 1942." To deal with war-related food shortages, animal husbandry classes and Victory gardens became commonplace in Los Angeles Unified schools. 1942.

But as the war ended, the intense focus on local food production receded. The Victory gardens and thousands of acres of farmland disappeared as Los Angeles entered an era of unprecedented growth. With expanding jobs in industry, as well as cheap home loans, especially for returning G.I.s, thousands came to Los Angeles. Throughout the 1950s, the county's population grew at a breakneck pace of more than sixteen thousand people per month, over five hundred newcomers every day. And as people settled down in Southern California, they wanted houses with yards, with schools and shopping close at hand. Postwar Los Angeles was set for the final transformation of its small farm suburbs to subdivisions and concrete.

First to fall to the bulldozers were vast acres of citrus groves spreading across the Los Angeles basin and its foothills. The iconic orange and lemon trees, central to the region's very identity, began to disappear, slowly at first, then, in the 1950s, with lightening speed. The papers were full of announcements for new subdivisions that would be developed over what was once fragrant citrus land, such as a 1951 Bellflower community, Gardendale Park, with almost two hundred homes "being fashioned from a former thirty-seven-acre citrus grove." The builders of Parkwood Homes, a 1951 Whittier development, wanted prospective buyers to know that "each lot has full-grown citrus trees, retained intact from the former citrus grove which stood on the property before it was subdivided." Even as they paved it over, developers of the new suburbs evoked an image of the land's abundance.

Freeway development, as well, cut great swaths through the citrus groves, like construction of the Santa Ana Freeway, which the *Times* described in 1950: "Where orange trees are growing one day, huge bulldozers are rumbling the next as highway construction crews rapidly push the Santa Ana Freeway more miles away from Los Angeles." From a peak of almost 4.5 million orange trees in 1945, Los Angeles County groves gave way to development, leaving less than a million trees by 1959.

Citrus, the county's most valuable harvest for years, was eclipsed in 1954 by nursery stock, a category of crops that includes trees, shrubs, flowers, and ornamental plants of all kinds. Always popular in garden-happy Los Angeles, ornamentals were selling like hotcakes, greening the landscapes of thousands of new homes and businesses. Even the expanding network of freeways needed landscaping, and lots of it, like the Hollywood Freeway, which was intended to "be an avenue of such loveliness as to excite the gardening world of the nation," according to the *Los Angeles Times*. City workers planted thousands of trees and plants, ranging from silk trees, to magnolias, to queen palms, to ivy and ice plants, alongside the

The Los Angeles County Fair grew out of a merchant's exposition held along the Southern Pacific Railway in downtown Pomona in 1921. The expo was such a success that the next year, local businessmen officially created the county's first fair to promote agriculture, horticulture, and animal husbandry. 1949.

A prize-winning bovine "smiles" for the camera at the 1951 Great Western Livestock Show. 1951.

freeway in 1951. Some five thousand trees were planted in just a two-and-a-half mile section from Silverlake Boulevard to Grand Avenue.

Even as its citrus groves and farms disappeared, Los Angeles at first hung on to its crown as queen of the nation's farm counties. In the USDA's 1949 farm census, L.A. still ranked as America's number one agricultural county. By 1954, though, it had fallen to third place, after Fresno and Kern Counties in California's Central Valley. Five years later, in 1959, it was in sixth place, trailing not just Fresno and Kern, but also Tulare and Imperial counties, as well as Arizona's Maricopa County. Between 1945 and 1959, L.A. County lost more than sixty percent of its farms, as the numbers dropped from more than 13,000 farms in 1945, to under five thousand in 1959. The amount of land in farms in the county fell more than forty percent between 1950 and the end of the decade, declining from 855,563 acres in 1950 to under a half-million in 1959.

The loss of farmland and open space seen during the postwar years in Los Angeles was in fact happening all over the U.S. *Fortune Magazine* called this trend "The Exploding Metropolis" in a 1957 series chronicling the rise of urban blight as farmland was paved over from Los Angeles to Chicago's expanding

An aerial view of orange groves near Covina shows how much agricultural acreage remained in Los Angeles County during the World War II era. 1941.

At the Los Angeles Union Stockyards, a buyer and seller negotiate. April 11, 1953.

suburbs, to communities around New York and Philadelphia that spread out and overtook small farms. The story of Los Angeles and how it lost its farms was not unique, but it was extreme.

Experts rang alarm bells about the overwhelming change taking place. William Whyte, one of the authors of the *Fortune* series, urged readers in 1958 to protect remaining open space in their communities, writing that "in the next three or four years, Americans will have a chance to decide how decent a place this country will be to live in, and for generations to come." These sentiments would help to spur the environmental movement in the coming decades.

About food production, though, there was less concern. The *Los Angeles Times* published a 1954 article on the decline of Los Angeles farming titled "Is Urbanization a Threat to the Food Supply?" The author concluded that given new technologies in production and distribution, there really wasn't much to worry about. It was true that technological advances were extraordinary after the war years, with improved varieties of crops, new machinery, chemical pest and weed control, more efficient fertilizers, and advancements in veterinary science. Farmers could now produce more milk from fewer cows, more eggs with fewer hens, and more fruit with fewer trees. And with the expansion of freeways, food could be trucked in faster, from further away. Still, the loss of farmland was worrisome, especially for those farmers who were struggling to continue.

As the 1950s progressed, the challenge to stay in business became overwhelming, especially as farm property was often taxed at the higher residential rate, instead of the agricultural rate. And increasingly, the neighbors weren't so fond of farming, finding it smelly and undesirable alongside brand-spanking-new communities. Dairies, which proved to be especially offensive to neighbors in shiny new subdivisions, were pushed out of most communities. By the early 1950s, many of the county's dairies had relocated to the southeastern corner of Los Angeles County, where, in 1956, politically active farmers created a specially zoned city, Dairy Valley, to provide a haven for farming. "We incorporated to prevent being zoned out of existence," said Dairy Valley mayor and local farmer Jaci Bettencourt. Dairy Valley, which later became Cerritos, housed more cows than people, with 32,000 cows but only 3,500 human residents in its nine square miles. At the end of the 1950s, Los Angeles continued its reign as the nation's number one dairy county, thanks to Dairy Valley's inventive zoning.

Even farmers who were determined to continue often found that their fight against the tide of development was unsustainable. In 1957, a Northridge walnut farmer, Art Record, who had owned his twenty-acre farm for thirty-six years, described to a *Times* reporter how hard it was to sell. "I almost know each tree by its name," he said. "I planted them all myself." He never planned to sell the farm where he and his wife had raised three daughters. But as the suburbs converged, his local walnut packinghouse closed due to lack of business, making it harder for him to market his crop. Then, as his property taxes began to skyrocket, up three hundred percent in just one year, the offers from developers became increasingly attractive. Selling, once unthinkable, began to seem inevitable. "My wife and I gave it a lot of thought," said Record. "And we decided to go along with the times." Throughout Southern California, thousands of farmers came to the same decision, as agriculture, once a reliable livelihood, became almost impossible in the face of the exploding metropolis.

Traffic stopped as sheep crossed Ventura Boulevard in Woodland Hills. According to the *Los Angeles Times*, traffic was backed up for a half-a-mile in each direction. 1955.

11 From Ag Land to Consumer Land, 1960 to 1990

Far from the sprawl of urban Los Angeles, the Antelope Valley has been home to outlaws, biker gangs, and even a socialist utopian community. In his 1990 book *City of Quartz*, Mike Davis called it "the last frontier of the Southern California Dream," a place where working-class families could still buy instead of rent. Encompassing more than one thousand square miles on the edge of the Mojave Desert, the Antelope Valley is a giant chunk of northern Los Angeles County, about the size of Rhode Island. Its two largest towns—Lancaster and Palmdale—fade away into empty stretches of desert, where occasional alfalfa fields and orchards dot the vast landscape of Joshua trees and tumbleweeds. Post-1960, farmers held on tenaciously in this last bastion of affordable land and slower urbanization, as the number of farms dwindled in the rest of the county.

After 1960, the Antelope Valley slowly became the agricultural center of Los Angeles County, if only because farming was fading in other parts of the county as it became urbanized. A 1974 County Crop Report touted the Valley's 20,000-acre record-breaking alfalfa crop, its expanding wheat and barley fields, and its profitable peach and pear orchards. Yet this was no agricultural paradise, like the L.A. basin had once been. Cold in the winter, blazing hot in the summer, relentlessly windy, and exceedingly dry, the High Desert challenged even the most experienced farmers. In 1973, the blowing soil from abandoned farmland contributed to dust clouds so massive they could be seen from space. Especially during a drought that lasted from 1985 to 1992, High Desert farmers survived on a razor's edge.

One Antelope Valley farmer, then president of the Los Angeles County Farm Bureau, expressed the

◄ Stampeding cattle escaped from an overturned truck on the Golden State Freeway, with humans to the rescue. 1981.

struggle of High Desert farmers to a *Los Angeles Times* reporter in 1987. Agriculture "has dropped most dramatically in the last five years," he said. "Some farmers just quit, others moved elsewhere, some went bankrupt, some sold to developers, and others began developing the land themselves. As I see it, in the next ten years most farmers will quit or die."

Reluctant to give up, farmers were always on the lookout for a crop that would bring a profit and help them stay in business. For a time, it seemed that a new peach variety might bring salvation to struggling Antelope Valley farmers. The particular peach was discovered in the 1970s by Jim Sprague, an enterprising High Desert resident who noticed that a peach tree in his yard, sprouted from a random pit, was producing delicious fruit. He propagated the tree and marketed it throughout the High Desert, touting his peach as especially tasty and productive. He called it the Last Chance, since it was the last peach of the season. At the same time, its name perfectly captured the sense that agriculture in the High Desert was running out of options. Some hoped that the Last Chance peach was a big opportunity to turn

An Antelope Valley peach orchard displays its spring colors. 2008.

The "control room" at Alta Dena's City of Industry dairy illustrates the advanced technological changes that shaped dairy farming in the latter years of the twentieth century. 1984.

things around.

But there are few miracle stories in farming. By the late 1980s, the rosy promise of success had faded from the Last Chance peach. One local farmer, who had spent two hundred thousand dollars planting over four thousand trees, was less than satisfied with the results, remarking, "If I had it all to do over, I'd keep my money in my hip pocket." If the peaches continued to disappoint him, he told a *Los Angeles Times* reporter, he would "subdivide the orchard for tract homes and name it Last Chance Acres." Despite such skeptics, Antelope Valley farmers planted more than three hundred thousand Last Chance trees in the 1980s.

Beyond the peach orchards, High Desert farmers grew the crops they had raised for decades, including alfalfa, onions, cherries, barley, and wheat. Some farms were small pick-your-own orchards, while others were large operations selling to national wholesalers. But it was a difficult business, especially for smaller farmers who increasingly took day jobs to supplement their farm income. Urbanization crept in, as suburbs sprouted up, even in this rural part of the county. Still, farming remained a visible and viable part of the economy in the Antelope Valley long after it faded in other parts of the county.

But while the High Desert was home to the lion's share of the county's remaining farm fields and

⮝ Hoping to draw support for the 12th Annual Forum Championship Rodeo, cowboys "drive" steers down Hollywood Boulevard. 1981.

orchards, farming was by no means gone from urban Los Angeles. In fact, because of an insatiable demand for plants to beautify the sprawling suburbs, an inedible form of agriculture generated the highest profits and even thrived in the face of population growth and skyrocketing land values: nurseries. Los Angeles area utilities, first Southern California Edison, then the Los Angeles Department of Water and Power, helped to boost the business of growing trees, shrubs, flowers, and other ornamental plants by offering land for lease under miles of power lines.

Other farm industries struggled, though, especially those deemed objectionable to local residents.

By the mid 1960s, even Dairy Valley lost its farming identity and became the modern, cow-free city of Cerritos, after it began allowing subdivisions. Farmers sold their land for huge profits, often moving east to dairy-friendly San Bernardino County. The number of dairies in the county fell dramatically, from almost three hundred dairies in 1966, to only seventy-five by 1972. By the end of the 1980s, just four commercial dairies survived in L.A. County.

"Pretty soon, you won't even see another cow around here. You'll only see a cow in movies," said John Vanderham, one of the county's handful of remaining dairy farmers in the late 1980s. Vanderham was

⮜ Roger Jessup's Dairy Farm in Pacoima was one of the county's largest dairies, with more than 1,200 cows on 125 acres at its peak. 1960.

▲ Twin boys cuddle Jerseymaid Holsteins brought to a Burbank Von's market to promote National Dairy Month. To educate urban dwellers about farming, California's farmers introduced public education programs like this one offered by the Dairy Council, and the Farm Bureau sponsored an "Ag in the Classroom" program. 1989.

frustrated with the City of Santa Fe Springs, which wanted him to move his cows back from the sidewalk and shield his dairy from public view. "All areas have to be screened so the people in the industrial area across the street don't look and see the mess out front," explained a city official. Agriculture, especially the kind that involved animals, and hence, manure, was viewed as a mess to be hidden, or better yet removed.

Flower and vegetable farming, once major L.A. industries, also dwindled. "Before World War II about ninety percent of the flowers we sold came from within the county," said Frank Kuwahara, the manager of the Southern California Flower Market in Los Angeles in a 1961 *Los Angeles*

▼ Patricia Baltazar harvests in a field that would eventually be part of the 105/Century Freeway. 1986.

Paul Hackney of the Newhall Land and Farming Company works in the company's citrus fields in the Newhall-Saugus area that would soon be bulldozed and developed. The company had 42,000 acres that were eventually turned into shopping centers, businesses, and housing tracts. 1961.

Times interview. "Now, it may not even be ten percent." Even that small percentage continued to decline, as farmers who had grown their flowers in Los Angeles County fields moved their production to less expensive land, north to Ventura County, or south to Orange and San Diego counties. A carnation or rose purchased from a local florist, which might once have been grown in L.A. County flower fields, was now more likely to be from a neighboring county, or more likely still, from a farm thousands of miles away in Guatemala or Costa Rica. By 1989, fewer than one hundred acres in the county were dedicated to producing cut flowers.

Acreage of the county's once abundant vegetable fields declined steeply as well, dropping by more than eighty percent in three decades, from more than 16,000 acres of vegetables harvested in 1960 to less than three thousand acres in 1990. And while fields of vegetable crops grew throughout the county in 1960, they were mostly concentrated in the Antelope Valley by 1990. Some formerly important crops, like asparagus and celery, were unsuited to High Desert heat, and slowly disappeared from the county's farms.

It wasn't just the high cost of land that drove flower and vegetable farmers to move their farms or abandon agriculture entirely. Urbanization was literally killing their crops, thanks to the pollutants in vehicle emissions. As the population grew, the ever-increasing number of cars spewed ozone and chemical compounds into the air, creating a smoggy haze over the Los Angeles basin. The smog stunted leaves, blighted petals, and slowed plant growth. Some especially sensitive crops became difficult to grow commercially in the county's polluted air, including leafy vegetables like spinach, as well as flowers such as chrysanthemums and orchids. "My spinach crop looked as if someone had passed over it with a blowtorch," Long Beach farmer Louis F. Martini Jr. told a *Los Angeles Times* reporter. The smog was just one more reason for farmers to move on.

Ironically, though, as agriculture disappeared throughout much of the

Lucio Estrada farmed along the Los Angeles River, a short distance from the Long Beach Freeway. 1989.

county, urban residents regained their enthusiasm for local, fresh food in the 1970s, reflecting the "back to the land" movement of the era. Enthusiastic backyard farmers planted gardens, raised poultry, and kept bees, especially on L.A.'s Westside. Although UCLA had shut down its agriculture department in 1960, partly in response to community demands that the campus farm be converted to additional parking, the university now offered community classes on backyard agriculture.

Even the Victory garden movement had a revival in this era when local food production again regained some importance. Mayor Tom Bradley started the Los Angeles Neighborhood Gardens and Farms Program in 1975 to convert vacant land into community gardens where families could grow their own food. Residents were at times ambivalent about this effort, with neighbors of a community garden in Reseda complaining to City Councilman Donald Lorenzen about "loss of privacy, fertilizer bags piled up against backyard fences, and odor and flies," according to the *Los Angeles Times*. Lorenzen used the issue to attack Bradley, claiming that the mayor's garden program was a cover-up and misuse of public funds, a dustup that the *Times* presented as a "Victory Garden War on Bradley." Despite its detractors, the program continued, supported by enthusiastic would-be gardeners around the city.

Mayor Tom Bradley and actor Eddie Albert, of television's *Green Acres*, "dig in" at the dedication of a community garden in Reseda. Any resident who wanted a plot at this, or future gardens, could apply at Los Angeles City Hall's Office of the Neighborhood Farm and Garden Project. 1975.

Los Angeles residents were enthusiastic about fresh produce, whether from a garden or a farm. In the late 1970s Angelenos began a love affair

Corn still grew in the Bunker Hill Community Garden in the late 1970s, even as buildings like the ultramodern Bonaventure Hotel stood just a short walk away. 1977.

with farmers markets that would grow in scope and influence in the coming decades. Before 1976, regulations made it difficult for farmers to sell directly to consumers, but modified rules sparked the first certified farmers markets in the state, including the first in Los Angeles County, which opened in the South Bay community of Gardena in 1979.

Farmers markets were a huge success with consumers and farmers alike. The *Los Angeles Times* described the hectic scene at the popular Santa Monica Farmers Market, which opened in 1981: "At precisely 11 a.m., the horn went off and the race was on. Idle spectators were wise to move aside as the Chinese eggplants, red and green bell and chili peppers and bags of juice oranges and grapefruits practically flew from the tables into the shoppers' waiting baskets, string bags, and wire carts."

Customers appreciated the variety of the produce and its just-picked flavor. For farmers, the markets represented a quick infusion of cash. "I probably would have lost my ranch if not for this," said a farmer who drove from San Luis Obispo County, some two hundred miles away, to sell his lettuce, spinach, and bell peppers at the busy Santa Monica market. Chefs, too, were fans of farmers markets, and upscale restaurants began to build their menus around purchases from their favorite farmers.

Along with tiny cooperative groceries and health food stores, farmers markets also supplied the small but growing segment of urban residents who wanted to buy food grown organically, without pesticides. With the publication of Rachel Carson's *Silent Spring* in 1962, consumers

Shoppers intently choose the freshest produce during the grand opening of the Santa Monica Farmers Market on July 31, 1981, a market destined to become one of the largest on the West Coast, frequented by shoppers as far away as Las Vegas. 1981.

became alarmed about the widespread use of agricultural chemicals like DDT, a concern that grew as the environmental movement picked up steam in the 1970s. First used during World War II, DDT, short for dichloro-diphenyl-trichloroethane, was long viewed as nearly miraculous. It controlled mosquitoes and a wide array of agricultural pests, with seemingly little impact on humans and animals. It was dusted and sprayed on farm fields, inside homes, and on city streets. The side effects, however, were far from miraculous. Birds, especially, were dying from exposure to the chemical, but until Carson sounded the alarm, its environmental impacts were overlooked. There were potentially serious consequences for human health as well. Carson evoked an image of a silent spring, where no birds sang, and her message hit home. Within a decade of her book's publication, the use of DDT was banned in the U.S.

Even so, the fallout from DDT was long lasting. Los Angeles County paid an especially high price for its rampant usage around the globe. In the late 1960s, scientists began to notice the disappearance of the brown pelican off the coast of Los Angeles. They identified DDT as the culprit, since it made pelican eggs so thin and delicate that they were crushed long before hatching. The pelicans, of course, had DDT in their system as a result of eating contaminated fish; further research proved that the water along the Southern California coast near Los Angeles had extraordinarily high levels of DDT. In fact, a federal study conducted in 1969 found that fish caught in the ocean off Los Angeles County had higher levels of DDT than fish caught off the Atlantic and Gulf Coasts, as well as other parts of the Pacific Coast. Where was all this DDT coming from, scientists wondered?

It didn't take long to determine the source. Los Angeles County was home to the world's largest DDT manufacturer, the Montrose Chemical Corporation, based in the city of Torrance. For twenty-four years, Montrose, which was responsible for two-thirds of the world's DDT sales, routinely flushed DDT into the county's sewers. Once in the sewers, it flowed into the ocean, leading to an

Agricultural workers Joanne Shimada and Rod Kerr sprayed Malathion in a Maywood neighborhood. 1987.

CalTrans workers installed Mexican fruit fly quarantine signs along the Long Beach Freeway in South Gate. 1983.

enormous deposit of toxic sludge two miles off the coast. Over one hundred tons of DDT, spread over seventeen square miles on the ocean floor off the Palos Verdes Peninsula, made it "one of the worst hazardous waste sites in the country," according to a U.S. Environmental Protection Agency official. In 1970, the Environmental Defense Fund brought suit against Montrose to halt their discharge of DDT, just the first of many lawsuits. Although the U.S. banned use of DDT in 1972, Montrose continued DDT production for a huge overseas market until 1983. Decades later, the massive DDT deposit still festers on the ocean floor off the coast of Los Angeles County. Fish caught in the area still have levels of the chemical that make them unsafe for human consumption.

Los Angeles area residents were further sensitized to the issue of agricultural pesticides in the 1980s in a high profile conflict over their use, when urban consumers faced off against the California Department of Food and Agriculture. At the heart of the dispute was the tiny Mediterranean fruit fly, or medfly, an imported pest which came to the continental U.S. in 1929, leading to aggressive eradication campaigns in Florida and Texas before it was first found in California. Devastating to agriculture, the medfly attacks more than three hundred types of vegetables, fruits, and nuts, many of which are important commercial crops in California, including citrus, apples, tomatoes, peaches, walnuts, and avocados. The adult flies lay eggs under the skin of fruit, where larvae soon hatch, turning fresh produce into a maggoty mess. Infestations have the potential to devastate California agriculture, since other states, along with major international trading partners restrict the entry of farm products from areas with the pest.

A procession of vehicles on Sunset Boulevard used loud speakers to protest Malathion spraying. 1984.

Unfortunately, with thousands of backyards full of host plants, a balmy climate, and sources of entry including an international airport, Los Angeles is a perfect place for the pest to thrive. And once it has a toehold in the backyards of urban and suburban residents, the fly can readily spread to farming areas. Los Angeles County became ground zero in a battle against the medfly with the 1975 discovery of California's first documented incidence of the pest in the westside community of Culver City.

Officials put a one hundred-square-mile quarantine in place, beginning an aggressive eradication campaign that featured ground spraying of the insecticide Malathion along with the release of sterile male medflies. Used to eradicate the pest in Florida and Texas, Malathion was considered safe for humans. The sterile flies, by mating with wild medflies, would help to reduce the population over time. The strategy seemed to work; by the following year, California officials declared victory, announcing the successful eradication of the medfly.

But the pest reappeared with a vengeance in 1980, detected in the San Fernando Valley community of Northridge on the same day that a larger Bay Area infestation was discovered. The State Department of Food and Agriculture went to war on the fly in an effort to protect California agriculture, but ground spraying did not eliminate the pest this time. At first, then-Governor Jerry Brown resisted a move toward aerial spraying, instead promoting an intensified campaign of ground spraying. Farmers, fearing for their livelihood, were furious with Brown for his delay of aerial spraying. The governor soon reversed his decision, after the federal government threatened to quarantine farm produce from the entire state. Beginning in 1981, Brown authorized aerial spraying of Malathion. Soon, residents of dense urban communities in the Bay Area and Los Angeles County endured late-night helicopter spraying.

Los Angeles County residents complained bitterly about the sticky mist sprayed by the helicopters, as well as their disturbing noise, described in a *Los Angeles Times* editorial. "There isn't much to see: four sets of red and green lights: then, as they come closer, their dark outline in staggered formation, four abreast, each one slightly behind the other. The sound is much

➤ Ida Honoroff led a protest against Malathion at the County Hall of Administration Building. Honoroff was a seventy-year-old consumer activist who wrote fiery speeches and pamphlets condemning the aerial Malathion spraying to control pests such as medfly and Mexican fruit fly. In one of her pamphlets she wrote: "One million people living in southeast and east Los Angeles County have become the targets of this undeclared war, with nerve gas poisons developed in Nazi Germany during World War II." 1984.

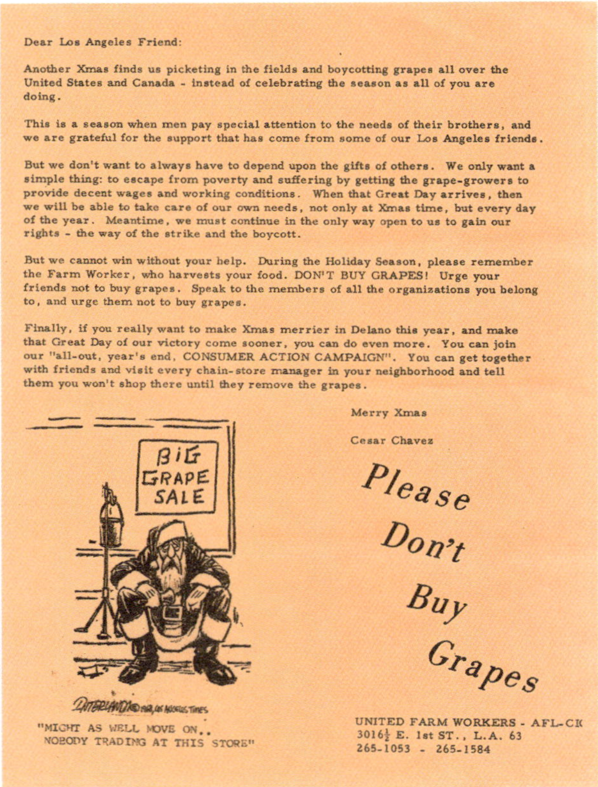

Dear Los Angeles Friend:

Another Xmas finds us picketing in the fields and boycotting grapes all over the United States and Canada - instead of celebrating the season as all of you are doing.

This is a season when men pay special attention to the needs of their brothers, and we are grateful for the support that has come from some of our Los Angeles friends.

But we don't want to always have to depend upon the gifts of others. We only want a simple thing: to escape from poverty and suffering by getting the grape-growers to provide decent wages and working conditions. When that Great Day arrives, then we will be able to take care of our own needs, not only at Xmas time, but every day of the year. Meantime, we must continue in the only way open to us to gain our rights - the way of the strike and the boycott.

But we cannot win without your help. During the Holiday Season, please remember the Farm Worker, who harvests your food. DON'T BUY GRAPES! Urge your friends not to buy grapes. Speak to the members of all the organizations you belong to, and urge them not to buy grapes.

Finally, if you really want to make Xmas merrier in Delano this year, and make that Great Day of our victory come sooner, you can do even more. You can join our "all-out, year's end, CONSUMER ACTION CAMPAIGN". You can get together with friends and visit every chain-store manager in your neighborhood and tell them you won't shop there until they remove the grapes.

Merry Xmas

Cesar Chavez

"MIGHT AS WELL MOVE ON.. NOBODY TRADING AT THIS STORE"

Please Don't Buy Grapes

UNITED FARM WORKERS - AFL-CIO
3016½ E. 1st ST., L.A. 63
265-1053 - 265-1584

An open letter from labor leader Cesar Chavez encouraged Los Angeles residents to boycott grapes in support of the United Farm Workers ongoing strike for better wages and working conditions. Although the center of the strike was to the north of Los Angeles in Kern County, Chavez said Los Angeles could influence the strike's outcome. Although it was no longer a production center, Los Angeles consumers still had the power to influence California agriculture. c. 1968.

more ominous. You hear it for a couple of minutes before you see them. It sounds like a Goodyear blimp flying low, loud and very fast. Everybody says the same thing: It feels like being in Vietnam." Some residents swore that the chemical was causing paint to peel off their cars. Others noted a lack of bees in their gardens and chalked it up to the spraying campaign. The controversy died down in 1982, when state officials announced that the pest had been eradicated. "This is a happy day in California as we declare the completion of the eradication of the last spot of infestation of medfly," said Richard Rominger, the state's director of food and agriculture, as officials drank a champagne toast to their victory.

Again, however, success was only temporary. A 1987 reemergence of the medfly in Los Angeles resulted in a new quarantine, along with extensive aerial spraying in the Los Angeles basin. To the dismay of residents, nighttime helicopter spraying of Malathion became common around Los Angeles County. Despite the continuing assurances of public officials that Malathion did not pose a health risk, urban dwellers were deeply suspicious of its effects. And there were now more dissenting voices among scientists about the safety of aerial spraying of the chemical over residential areas.

By 1990, a number of cities in Los Angeles County sued the State of California to halt the spraying. Protests were common as the national media focused on the controversy surrounding spraying in Los Angeles. Dissent, L.A. style, sometimes included celebrities, such as a 1990 anti-spraying protest in West L.A. led by cast members of the television show *Cheers*. In late 1990, the state once more declared success, though victory would yet again prove elusive.

The ongoing medfly battle highlighted a new reality. Although Los Angeles was no longer a major farming area, it remained central to the success of California agriculture. Agriculture and urban communities were more separate geographically, yet increasingly interdependent economically. However, it was an uneasy coexistence. Urban residents often worried about how their food was produced, while farmers fretted that city dwellers had little understanding of agriculture. The American Farm Bureau Federation, concerned about consumer attitudes toward agriculture throughout the nation, conducted a 1990 survey that showed eroding public trust in farmers, especially in regard to protecting the environment and ensuring food safety. Sixty-nine percent of those who responded thought that farmers were too heavily influenced by chemical companies in their use of farm chemicals, while more than seventy percent "agreed that farmers are more interested in surviving economically than they are about using chemicals safely."

Farmers expressed their frustration with public attitudes toward agriculture at a Los Angeles meeting in late 1990. "It's misinformation that they're basing their consensus on now...It just doesn't make sense," commented a retired Los Angeles County onion farmer, head of the Los Angeles County Farm Bureau, to a *Los Angeles Times* reporter. Another farmer remarked; "We've all been touched by this loss of consumer confidence. I liken this whole affair to some uneasy wakeup call." Farmers throughout the state responded to the public's unease through participation in education efforts such as the Farm Bureau-sponsored Ag in the Classroom program, which aims to teach young people about the importance of California agriculture. And, increasingly, they worked to meet the growing demands of the public for food that was produced organically or with fewer pesticides.

As Los Angeles became an epicenter of consumer power, wielding influence over California's farmers, it was no longer a major agricultural county, though plenty of farms still existed. By the 1990s, its agriculture was increasingly an invisible industry, pushed out beyond the urban core to the wide-open spaces of the desert, or hidden under power lines and in utility right-of-ways. Most Los Angeles County residents did not know that their homes had once been part of an agricultural empire, where fruit, flowers, and vegetables had grown in abundance. Los Angeles County's farm history became a forgotten story, paved over like the countless fields and orchards that disappeared under the spreading suburbs, a past receding as if it had never existed at all.

Epilogue

North of Magic Mountain and its glittering roller coasters, Route 138 travels east from the Golden State Freeway toward the wide-open spaces of the Antelope Valley. With few people, little business, and only a handful of cars, the High Desert does not feel like Los Angeles, at least not the one most people know or imagine.

In 2013, signs along this rural road advertised pick-your-own peaches, such as the sign leading to the M&M Peach Ranch, a picturesque orchard set against a backdrop of mountains in the tiny community of Lake Hughes. On a Saturday in September, despite acres of trees loaded with ripe fruit, the ranch was mostly empty. Sixty-odd miles from downtown L.A., perhaps the M&M was just too far away for even the most enthusiastic foodies and locavores. They didn't know what they were missing, though: the ranch's sweet, juicy Last Chance peaches were in season. The M&M was one of the last commercial orchards in the county focused on the Last Chance peach, L.A. County's once popular homegrown variety.

This farm, like others that continued after 1990, was still bountiful, often struggling, on the far edge of the Los Angeles metropolitan sprawl. Most Los Angeles County residents were unlikely to see the farms sprinkled around the High Desert, a little-visited part of the county. Angelenos were more likely to catch a glimpse of hogs or cattle packed tight in tractor-trailers, or trucks piled high with tomatoes and other crops, rumbling past on the freeway as they made their way from California's farm fields toward market.

Once a production center, Los Angeles was still a marketing and transit center. In its 2007 County Crop and Livestock Report, the Los Angeles County Department of the Agricultural Commissioner

◄ Ron Finley, a South Los Angeles resident, achieved celebrity status after he and his "illegal" edible parkway garden were prominently featured in the media. 2014

called it "the great conduit of California agriculture." Despite the disappearance of local farms, according to the report: "We're still an agricultural hub. The Ports of Los Angeles and Long Beach are an unparalleled American import/export gateway. Los Angeles International Airport handles enormous volumes of cargo. The Los Angeles Wholesale Produce Market, the largest of its kind in the nation, bustles with customers in the heart of Los Angeles." The massive produce market spreads over several blocks on the edge of downtown L.A. Hours before dawn, while the rest of the city sleeps, trucks unload at the market's vast warehouses. Buyers from restaurants and grocery chains arrive to make their purchases, from mangoes by the case to lettuce by the pallet. Much like farming, this was a largely hidden segment of L.A.'s economy.

But while it became more important as a hub for the transit of agricultural goods, Los Angeles County was still a significant producer after 1990. In fact, farming had something of a renaissance due to one particular crop. Fields of carrots, with their distinctive feathery, bright-green leaves, popped up around the Antelope Valley where alfalfa, barley, and orchards once grew. Busy consumers who wanted healthy convenience foods drove the growth of the carrot industry as baby carrots became popular. Bags of ready-to-eat baby carrots, which are actually regular-sized carrots peeled and cut into uniform "baby" proportions, are a favorite of parents, dieters, and healthy eaters. Pioneered in the 1980s, then popularized in the '90s, baby carrots transformed the carrot industry. By 1999, American consumers preferred them to whole carrots, crunching on thirty-three percent more carrots than they had at the beginning of the decade.

As demand for carrots soared, giant growers who operated out of Bakersfield looked around for new carrot-farming territory. They soon came to L.A. County's High Desert, which has the kind of sandy soil that allows carrots to grow long and straight. Before long, Los Angeles County was a major player in carrot farming. Surprisingly, according to the USDA's 2012 Census of Agriculture, conducted every five years, Los Angeles County was the fourth largest carrot-producing county not only in California, but the entire nation. Previously, the category of root vegetables in the annual Los Angeles County crop report, which includes turnips, beets, and carrots, had all but disappeared, with only seven acres and $30,500 in revenue reported in 1992. As the demand for carrots picked up steam, though, root vegetables made a remarkable comeback. By 2002, there were almost ten thousand acres of carrots, with more than $44 million in sales. For some Antelope Valley farmers, this new crop was all that allowed them to continue. "Carrots have kept me in business. I'm bringing on three hundred new acres a year," said one farmer, interviewed by the *Los Angeles Times* in 2001.

But even with the booming demand for carrots, there was a crippling challenge. Water, always the Achilles heel of High Desert farming, resurfaced as a critical issue. For decades, water pumped from underground aquifers helped to make farming possible in the hot, dry Antelope Valley. This water also made a housing boom possible, with new developments spreading throughout the High Desert in the 1980s and '90s. Since both communities and farms depended on inexpensive underground water, it became clear that with so much growth in the Antelope Valley, groundwater was simply being used up faster than it could be replaced by melting snow and runoff from the mountains.

Farmers maintained that they owned the water under their land, that their rights should not be compromised by the sprawl of new housing tracts. In particular, influential carrot growers wanted to secure the use of water under their land to ensure the future of their operations. In 1999, carrot grower Diamond Farming Co. sued local water districts, alleging that they were slurping up the water underneath Diamond's land, providing it to other users. Before long, more carrot growers and other farmers joined the fray, entering the complicated twilight zone of California water law.

California, as the Antelope Valley growers found, was the only Western state that lacked comprehensive regulations around groundwater use. This meant that there were no easy answers about who had the rights to groundwater. When there was a dispute, the courts attempted to sort out competing claims through a lengthy process called adjudication, which often took years. A decade and a half after it started, the Antelope Valley case was still not resolved.

Aside from the water issue, development also pushed up against farming in the Antelope Valley. More often than not, agriculture lost out to subdivisions. As farming declined in the Antelope Valley, the numbers reflected the continuing erosion of L.A. County's farm heritage. Farms in the county fell from 1,446 farms in 1992 to 1,294 in 2012, which doesn't seem to be a huge decline over the course of twenty years. But acreage tells a fuller story. In the same two decades, the acres of Los Angeles County land in farms fell by half, from 183,569 to 91,689 acres. That was a whole lot less land being farmed.

While the ongoing disappearance of farmland around the county went largely unnoticed, the popularity of urban agriculture was a more visible trend. A number of projects, including community gardens and mini-farms in the city were high profile, harkening back to WWII Victory gardens, as well as the neighborhood garden program started by Mayor Tom Bradley in the 1970s. Often, public health officials, gardeners, activists, and politicians looked to urban gardens and farms as part of the solution for food insecurity, a problem that plagued Los Angeles, especially after the economic crisis of 2008.

"Food insecurity is a major public health issue that has reached crisis proportions in L.A. County," according to a 2015 Los Angeles County Public Health Department report. "It can have significant negative impacts on health and well-being across the lifespan, including impairing growth and development among children, increasing risks for depression and other mental health conditions among adolescents, and contributing to malnutrition and worsened medical conditions in the elderly." Food-insecure families are unable to consistently afford adequate healthy food, and many skip meals to make it through the month. In many cases, food-insecure families experience hunger on a regular basis. Food insecurity, a crisis associated with poverty, was a daily reality for almost 1.5 million residents of Los Angeles County in 2015. Tight budgets mean more cheap food, which is often less healthy food. Crunchy apples, crisp lettuce, and sweet strawberries are a pricey luxury, while burgers, fries, and chicken nuggets are on the dollar menu, a diet that drives obesity, diabetes, and heart disease.

What's the solution to food insecurity and its related health issues? A big part of a healthy diet and food security is having plenty of fresh produce with every meal. "A diet rich in vegetables and fruits can lower blood pressure, reduce risk of heart disease and stroke, prevent some types of cancer, lower risk of eye and digestive problems, and have a positive effect upon blood sugar which can help keep appetite in check," according to the Harvard School of Public Health.

The importance of produce was reflected in "MyPlate," the USDA graphic introduced in 2011 that represents a healthy meal, which shows that fruits and vegetables should make up half of every plate of food. Yet for many Angelenos, this recommendation was out of reach. Fresh produce was an occasional treat, rather than a foundation of their diet. Access is a problem as well as affordability, since L.A.'s inner city has long had far fewer grocery stores than its suburbs. That's where gardens come into play. In a place where poverty is entrenched, with fruits and vegetables neither affordable nor accessible, why not use backyards, vacant lots, and other open spaces to grow healthy food, right in the neighborhoods that need them?

That's what Ron Finley thought when he planted vegetables in a parkway median in front of his South Los Angeles home in 2010. In the City of Angels, the residents are required to tend to this narrow strip between the sidewalk and the street. To Finley, it seemed like the perfect spot to grow vegetables for his food desert community. He transformed the ten-foot wide stretch of patchy grass into a thriving garden full of chard, onions, peppers, and tomatoes, encouraging his neighbors to share in the harvest.

The sidewalk garden seemed to be a simple, practical solution in a food-insecure community, but Finley was soon slapped with a citation. According to city law, only a certain selection of non-edible plants could be grown in parkway strips, so his garden was illegal. Remove the garden, city officials said, or pay several hundred dollars for a special permit. Finley did neither, and instead took the issue to City Hall, working with community activists to change the rules. Given the city's bureaucracy, where officials worried about slip and fall lawsuits caused by errant sidewalk tomatoes, changing the law was not easy, but, after a four-year fight, Finley prevailed. In March 2015, the City of Los Angeles made it legal to plant edible crops in parkways.

Finley and his edible parkway became a cause célèbre, widely featured in the media, not only in local media like the *Los Angeles Times*, but also the *Wall Street Journal*, the *New York Times*, *CNN*, and even Southwest Airlines *Spirit Magazine*, with much of this attention driven by Finley's wildly successful 2013 *TED Talk*. In his talk, Finley shared the visible impacts of the modern food system in his community. "I see wheelchairs bought and sold like used cars. I see dialysis centers popping up like Starbucks. And I figured, this has to stop. So I figured that the problem is the solution. Food is the problem and food is the solution." He described the power of gardening to transform communities. "Gardening is the most therapeutic and defiant act you can do, especially in the inner city. Plus, you get strawberries." The simplicity of his solution resonated with his audience.

However, accessing land to grow strawberries or anything else is not easy in Los Angeles. In 2013, according to a Harvard study, the percentage of residents who rented rather than owned a home in the Los Angeles metropolitan area, at fifty-two percent, was the highest in the country. In L.A.'s dense urban neighborhoods, apartment dwellers might have a balcony, but probably not a backyard with space for a garden. As difficult as it was for Ron Finley to secure the right to grow food on the tiny patch of land between the sidewalk and the street, for many Angelenos, obtaining land for gardening is even harder. This is despite the fact that more than eight thousand vacant lots were documented in the City of Los Angeles in 2014, many of them concentrated in food desert communities like South Los Angeles.

An infamous fourteen-acre plot of land in another South L.A. neighborhood provides a case in point. In 2016, the land at Forty-first and Alameda Streets sat empty behind a chain-link fence, its scraggly weeds and patches of bare dirt showing little evidence of its former abundance. Ten years earlier, it was a lush oasis of tomatoes, peppers, cilantro, corn, and fruit trees, where more than 350 families grew food and found community. Started in 1994, the South Central Garden was the largest community garden in the United States, a bountiful source of food for the surrounding low-income neighborhood, where there were few grocery stores, but no shortage of potential gardeners.

Yet despite its success, the garden came to a dramatic end after a lengthy and high-profile struggle between the gardeners and the land's new owner. "It's complicated" would describe the relationship of the land to its chain of custodians. For the first nine years of the garden's existence, the site was owned by the City of Los Angeles, which had purchased the parcel through eminent domain in the 1980s, planning to build a giant trash incinerator on the property. When the community organized against the incinerator, that plan was abandoned, and the land sat empty until it became home to the flourishing garden.

Fourteen acres is a lot of land in urban Los Angeles. Developers eyed the lot with interest, especially since it was along an important transportation corridor. In 2003, the site was sold in a closed-door session, when city officials allowed developer Ralph Horowitz, a former owner of the site, to reclaim the land at a price far below its market value. The gardeners, who leased the land, were asked to leave, although they delayed their eviction through a legal case questioning the legitimacy of the city's deal with Horowitz. Ultimately, they lost the case along with their garden. In 2006, sheriff's deputies forcibly removed gardeners and protesters, bulldozing the garden while families who had tended fruit trees and plots of vegetables watched as their labor of love was destroyed in moments.

The struggle for the land was high profile, with coverage in national media, and an Academy Award-nominated documentary, *The Garden*, released in 2008. But even with media attention, involvement of celebrities like Daryl Hannah and Willie Nelson, and a successful fundraising campaign, the mostly low-income, immigrant gardeners could not prevail in a situation where the political deck was stacked against them. While still empty in 2016, the land had yet another owner, a conglomerate of clothing manufacturers with plans to construct four large warehouses on the site. The convoluted history of the land at Forty-first and Alameda, as well as the outcome concerning its use, seemed emblematic of land-use politics in Los Angeles, where development has often obscured and obliterated other community interests.

For those Angelenos lucky enough to have a backyard, it was much easier to raise food. The popularity of home gardening was a well-documented trend, at least at the national level. As of 2013, one in three U.S. households grew at least some of their own vegetables, fruits, and culinary herbs, according to the National Gardening Association (NGA). Edible gardening was on the rise: between 2008 and 2013, there was a seventeen percent increase in Americans growing their own fruits and vegetables, whether in a backyard, a patio container, or a community garden plot. And in urban areas, this trend was even stronger, up twenty-nine percent in that same five-year period. According to the study, gardeners grew their own food because they thought it tasted better, believed that they were saving money, and considered the food they grew to be higher quality than grocery store produce.

In addition, according to the NGA, many gardeners believed that their homegrown fruits and vegetables were safer than commercially grown produce. Scientists, farmers, and consumers hotly debated the validity of such beliefs, but many consumers distrust corporate agriculture, which feeds the fears. This unease, which sprouted in the 1970s and '80s, continued to grow, fueled by documentaries like 2008's *Food, Inc.*, and books such as *The Omnivore's Dilemma*. Whether it was concerns about pesticide residues or GMOs (genetically modified organisms), many Angelenos, like other Americans, worried about the food at the grocery store. In 2015, according to a USDA report, more farmers were selling directly to food-conscious consumers, bypassing the network of middlemen and markets to reach customers who want to know more about how their food is grown. This was reflected in the nationwide increase in the numbers of farmers markets, which rose seventy-six percent between 2008 and 2014. If the crowds at popular local farmers markets like the Hollywood and the Santa Monica markets were any indication, Angelenos heartily supported small farmers and local food.

The passion for local food was also manifested in the number of Angelenos who went beyond raising a garden, embracing traditions straight out of L.A.'s history of backyard mini-farms and small farm homes. One such activity was beekeeping, a throwback to a time in the late 1800s when "bee ranches" dotted the hills around Los Angeles. After 2010, beekeeping became high profile in Los Angeles, popular among Angelenos who were not content to buy local honey at a farmers market, but wanted their own honey, from their own bees, in their own backyard.

Unfortunately, backyard beekeeping was not exactly legal in many parts of Los Angeles County. There was a confusing mix of beekeeping rules in Los Angeles County, which varied from city to city, based on complex municipal codes. Bees were legal in the City of Santa Monica beginning in 2011, for example, while next door in the City of Los Angeles, they were prohibited until 2015, when beekeeping activists successfully convinced L.A. officials to legalize beekeeping, in keeping with other major U.S. cities, such as New York, Chicago, and San Francisco. They often encountered intense opinions on urban beekeeping, both pro and con.

Detractors argued that bees have no place in the city, including one Los Angeles city council member who remarked, "Why would you bring that dangerous scenario to someone's backyard? There are health issues. There are danger issues. It shouldn't be in a residential neighborhood. There are too many instances where innocent bystanders can be the victim."

On the contrary, according to urban beekeeping activists, it's common sense to legalize backyard bees. Bees are already present in the urban environment, flying across artificial city and zoning boundaries, whether we like it or not. And with honeybees on the decline around the world due to the mysterious Colony Collapse Disorder, some beekeepers believe that raising bees in the city is a way to boost the population of these critical pollinators. Indeed, some research showed that city bees were faring better than rural bees, thanks to the diversity of flowers in urban areas that provide abundant food for bees. With a quarter of our diet comprised of crops pollinated by bees, including apples, almonds, avocados, cucumbers, melons, squash, peaches, and pears, to name just a few, the honeybee crisis struck a chord with local residents.

Backyard chickens are another trend that both resonated with and generated conflicting opinions among urban Angelenos. Once a popular business among L.A. residents trying to make a buck from their small farm homes, chickens have experienced a surge of popularity in urban and suburban neighborhoods. "I've lived here for 50 years. All of a sudden there's an influx of chickens. You're not supposed to have chickens in the city," a retired librarian, living near Watts, told a *Los Angeles Times* reporter in 2008.

Community annoyance over crowing roosters grew loud enough that the City of Los Angeles passed a 2009 ordinance to limit the number of roosters to only one per household.

On the other hand, in some neighborhoods, backyard poultry hobbyists invest in heritage breeds and fancy hen houses. One chicken-keeper in L.A.'s upscale Silver Lake wrote an account of her experience as a new poultry owner for the *L.A. Times Magazine*: "It took three months of weekends to build a veritable chicken palace. It was a sight to behold, complete with insulation, rain gutters and, yes, a motorized door on a timer. There is an outdoor run that no animal other than a bear could breach. I have lived in crappier apartments. We were ready—bring on the birds!"

In fact, in 2011, the birds were so popular that *Sunset Magazine* named backyard chickens one of the "Top 100 Cultural Trends in the West." So why did chickens become a social phenomenon? It's not entirely clear, but like the home-gardening trend, may be rooted in concern about the quality and safety of food, as well as economics. Gardening took off after the economic crisis of 2008, with backyard chickens and beekeeping following the same trajectory. Given lingering concerns about economic stability, perhaps a backyard garden, a small flock of laying hens, or honey from one's own hive, helped make life feel a little less precarious. Backyard farmers often strive for a more sustainable lifestyle, one where they can eat out of their garden and save on their groceries, although whether these things actually save money is an open question. For example, while many backyard poultry enthusiasts believe they are pinching pennies, in 2015, *U.S. News and World Report's* "My Money" columnist saw it differently: "A number of backyard egg advocates have crunched the numbers—and sadly, this is not a money-saving endeavor for most." But money-saving or not, the home food production trend continued in 2015.

The consumer appetite for food perceived as healthy and sustainable—whether grown in a backyard, or purchased at a farmers market or grocery store—was impacted by California's water problems. Long-term drought inevitably increases prices of everything— from fruits and vegetables, to eggs and milk. And a couple of rainy winters don't solve things. In Southern California, the ongoing dispute over use of groundwater in Los Angeles County's High Desert that began in 1999 foreshadowed events in the rest of the state. In 2014 and 2015, in the grip of an epic drought, California farmers grappled with rapidly dwindling water resources. Fights over water, nothing new in California, gained a gritty air of desperation as crops and orchards shriveled around the state.

In 2014, with scant surface water available for farms, farmers dug deeper wells, keeping the state's well drillers busy around the clock. Previously ignored and neglected, groundwater was fully in the limelight that year, with Governor Jerry Brown pushing to finally regulate California's underground water resources. The governor signed legislation in September 2014 that required slow, cautious steps toward protecting the state's aquifers, but water experts feared that the measures were too little, too late. The groundwater crisis called into question the long-term viability of California's farms, both large and small. Backyard gardens, too, were challenged by historic mandatory cutbacks in municipal water use implemented in April 2015.

Drought was not the only specter haunting California agriculture and its future success in 2015. The potential for devastation of California crops by invasive pests continued to emerge during Brown's second life as California governor, just as it did in his first. His old adversary, the medfly, had taken up permanent residence in California. After three decades of efforts to eradicate the pest, many scientists came to consider it endemic, a problem to be managed rather than eliminated. And other invasive pests followed the medfly in the pantheon of threats to California's agriculture. Quite a few, like the medfly, first cropped up in Los Angeles County, which has often proven to be invasive-pest central.

One major threat, delivered to California agriculture by way of Los Angeles, is huanglongbing, or citrus greening, a bacterial disease originating in China, with the potential to destroy California's multi-billion dollar citrus crop. A tiny insect called the Asian citrus psyllid carries the disease, injecting bacteria into citrus trees when the infected organism feeds on their leaves. Soon, the leaves of a formerly healthy orange or lemon tree begin to yellow, its fruit become misshapen, and its juice bitter. Within a handful of years, the tree will die.

California's farmers had long dreaded the arrival of the psyllid and the disease, which devastated the citrus industries in Florida, China, and Brazil. Despite efforts to keep the problematic insect out of the Golden State, it was detected in San Diego in 2008, then in Orange County and the Los Angeles community of Echo Park in 2009. Despite quarantines, the pest soon spread to other counties, but there were no reported incidences of huanglongbing. California citrus growers held their collective breath. Their fears were realized in 2012, when the disease was identified in a tree in Hacienda Heights, a community in eastern Los Angeles County.

Situated in a lush backyard orchard, the tree in question was a lemon tree with a grafted branch of pomelo, a fruit similar to a large, mild grapefruit with an extra-thick rind. And this grafted branch was the source of the disease, arriving through the most innocent of activities. Homeowner Mary Wang and her friends from church shared a love of gardening, often trading seeds, plants, and cuttings. A friend from this group, who helped her with her backyard orchard, grafted the pomelo branch onto the lemon so that the tree would produce two kinds of fruit. The pomelo, a symbol of good fortune and prosperity in Chinese culture, was obtained in a roundabout way, traded from friend to friend, its original source unknown.

Ironically, Los Angeles County, the cradle of citrus culture in California, was now the potential source of its destruction. From its launching pad in a single Los Angeles County backyard, the disease was expected to move to other parts of the state. With no cure for the disease, the focus shifted to controlling the Asian citrus psyllid, using everything from pesticides to the release of tiny psyllid-eating wasps from Pakistan in the arsenal to control the psyllid. In 2015, with pockets of the pest established around the state, the spread of the disease was only a matter of time, putting the future of California's iconic crop in jeopardy.

Loss of farmland, too, threatened California agriculture. The story of agriculture in Los Angeles, and how it lost its farmland, was also the story of California. Just as thousands of acres of prime farmland were lost to development in Los Angeles County as it urbanized, the same has happened around the state. According to the California Department of Conservation, between 2010 and 2012, the state lost more than 58,000 acres of farmland, an area larger than the city of Riverside, to non-farming uses, which often meant that productive land was permanently lost under homes, shopping centers, businesses, freeways, and parking lots. Los Angeles County paved over its farmland faster, on a grander scale, but other counties around the state have mostly followed the same path.

The triple threat of drought, invasive pests, and loss of farmland called into question the future of the Golden State's farm bounty. Yet despite the fragility of our food system, the state's population continued to swell. Would California, the nation's top farm state, lose its productive capacity until fruits and vegetables, milk, and other food must come to our plates from other places, because we can no longer produce them ourselves? Look no farther than Los Angeles for a cautionary tale. Consider this statistic from a

1940 Chamber of Commerce publication: "Nearly half of the Los Angeles food supply originates on farms within fifty miles of the city." In 2015, it was hard to say how much it would be, except that it was probably a tiny fraction of a single percent.

Farm history in Los Angeles, with its epic rise and fall, is not unique; it is the past of California itself. California, like Los Angeles, was a place of abundant natural resources, with varied food sources gathered by the native people who lived here for centuries. It was a land where newcomers became grain producers and cattle ranchers, slowly converting a remote outpost into a population center; a place where a modern farming empire rose up, built on irrigation, technology, and cheap labor, its fruits, vegetables, nuts, milk, cotton, beef, eggs, and honey bountiful enough to export throughout the world. And ultimately, a place where development was valued more highly than farms, speeding the transition from agrarian to urban, from cows to concrete.

The story of agriculture in California is still unfolding. What happens will surely depend on how the state handles its drought crisis and manages the related wildcard of a warming climate. The future of farming in Los Angeles County, in particular, seemed difficult in many ways, but there was still hope for agriculture in the nation's most populous county. A 2013 study conducted by UCLA students, titled *Cultivate L.A.*, documented more than 1,200 urban agriculture sites in the county, which they defined as including school gardens, community gardens, nurseries, and farms. In 2015, officials in both the City and County of Los Angeles intent on boosting these numbers, made plans to give property owners a tax break for offering vacant land for use as farms or community gardens.

Los Angeles farmers have always been creative in finding ways to beat the odds. In the second decade of the twenty-first century, urban farmers around Los Angeles experimented with novel approaches that could potentially make farming in the city profitable, from vertical growing to indoor farms to rooftop agriculture. One example was Local Roots Farms, a start-up growing lettuce inside shipping containers using hydroponics, a system that uses significantly less water than conventional systems to grow fresh produce. With new farming methods and supportive policies, farms and gardens may continue to thrive though they will face many challenges.

Hints of the possibilities sometimes came from unexpected places. In 2015, new tenants reinvigorated M&M Peach Ranch, the orchard that seemed so empty just two years earlier. Some of the gardeners from the South Central Garden, who refused to accept defeat after their high-profile battle ended in 2006, formed South Central Farmers, a successful fruit and vegetable enterprise located 130 miles north of Los Angeles in Kern County. In 2015, when they wanted to expand, the farmers connected with Bill Basner at M&M Ranch, who had acres to spare. They planted vegetables at M&M, which they featured in baskets of produce offered on a sliding scale to customers throughout urban Los Angeles. Like other farming in the Antelope Valley, the success of the union hinged on access to affordable water. But in the meantime, it was a place to grow, where the long tradition of Los Angeles County's abundance was sustained. From the fields of the Antelope Valley, to urban backyards and community gardens, the harvest continued in Los Angeles County, once an agricultural empire, still a land of possibilities and reinvention.

A celery field in Venice. Venice and the surrounding area produced so much celery each year that it was referred to as the Venice Celery District, and even the "Celery Capital of the World." Fields of celery were grown in the Celery District as late as the 1970s, in the area around Lincoln and Jefferson Boulevard, where Playa Vista sits. 1927.

Selected Bibliography

This bibliography is divided into sections. First is a list of key resources that informed multiple chapters. These works helped to shape our understanding of the history of Los Angeles and California, including the development of agriculture. The books, articles, and pamphlets listed here were consulted frequently throughout the project. Following these general resources, chapter-specific references are noted that provided additional information as well as important facts and details for each chapter. All citations to the *Los Angeles Times* newspapers are referenced as *LAT*. We garnered agricultural statistics from the annual Crop and Livestock Reports produced by the Los Angeles County Department of the Agricultural Commissioner, from the California Department of Food and Agriculture, and from the U.S. Department of Agriculture's Agricultural Census. —R.S. and J.G.

Key Resources:

1. Anderson, M. Kat. *Tending the Wild: Native American Knowledge and the Management of California's Natural Resources.* Berkeley: University of California Press, 2005.
2. Barraclough, Laura R. *Making the San Fernando Valley: Rural Landscapes, Urban Development, and White Privilege.* Geographies of Justice and Social Transformation. Athens: University of Georgia Press, 2011.
3. Beebe, Rose M., and Robert M. Senkewicz, eds. *Lands of Promise and Despair: Chronicles of Early California, 1535–1846.* Santa Clara, California: Santa Clara University, 2001.
4. Carosso, Vincent P. *The California Wine Industry, 1830–1895: A Study of the Formative Years*. Berkeley: University of California Press, 1951.
5. Chambers, Clarke A. *California Farm Organizations: A Historical Study of the Grange, the Farm Bureau, and the Associated Farmers, 1929–1941*. Berkeley: University of California Press, 1952.
6. Chan, Sucheng. *This Bittersweet Soil: The Chinese in California Agriculture, 1860–1910*. Berkeley: University of California Press, 1986.
7. Cleland, Robert Glass. *The Cattle on a Thousand Hills: Southern California, 1850–80*. 2nd ed. San Marino, California: The Huntington Library, 1941.
8. Davis, Mike. *City of Quartz: Excavating the Future in Los Angeles*. 2nd ed. New York: Verso, 2006.
9. Deverell, William, and Greg Hise, eds. *Land of Sunshine: An Environmental History of Metropolitan Los Angeles*. Pittsburgh: University of Pittsburgh Press, 2005.
10. Dumke, Glenn S. *The Boom of the Eighties in Southern California*. San Marino, California: Huntington Library, 1944.
11. Estrada, William David. *The Los Angeles Plaza: Sacred and Contested Space*. Austin: University of Texas Press, 2008.

12. Fogelson, Robert M. *The Fragmented Metropolis: Los Angeles, 1850–1930*. Berkeley: University of California Press, 1993.
13. "From Agriculture to Aerospace." Los Angeles: Los Angeles Area Chamber of Commerce, 1988.
14. García, Matt. *A World of Its Own: Race, Labor, and Citrus in the Making of Greater Los Angeles, 1900–1970*. Studies in Rural Culture. Chapel Hill: University of North Carolina Press, 2001.
15. Gilbert, Jess, and Kevin Wehr. "Dairy Industrialization in the First Place: Urbanization, Immigration, and Political Economy in Los Angeles County, 1920–1970." *Rural Sociology* 68, no. 4 (Dec. 2003).
16. Gillingham, Robert C. *The Rancho San Pedro: The Story of a Famous Rancho in Los Angeles County and Its Owners the Dominguez Family*. Los Angeles: The Dominguez Properties, 1961.
17. Gonzalez, Gilbert G. *Labor and Community: Mexican Citrus Worker Villages in a Southern California County, 1900–1950*. Statue of Liberty-Ellis Island Centennial Series. Urbana: University of Illinois Press, 1994.
18. Guinn, J.M. *A History of California and an Extended History of Its Southern Coast Counties, Volume I*. Los Angeles: California Historic Record Co., 1907.
19. Gumprecht, Blake. *The Los Angeles River: Its Life, Death, and Possible Rebirth*. Creating the North American Landscape. Baltimore: Johns Hopkins University Press, 1999.
20. Hamilton, Andrew, and John B. Jackson. *UCLA on the Move: During Fifty Golden Years 1919–1969*. Los Angeles: Ward Ritchie Press, 1969.
21. Jackson, Robert H., and Edward Castillo. *Indians, Franciscans and Spanish Colonization: The Impact of the Mission System on California Indians*. Albuquerque, New Mexico: University of New Mexico Press, 1996.
22. Johnston, Warren E., and Alex F. McCalla. "A Stylized History of California Agriculture from 1769 to 2000." In *Whither California Agriculture: Up, Down or Out? Some Thoughts About the Future*. Davis, California: Giannini Foundation, University of California, 2004.
23. Los Angeles Chamber of Commerce. *Los Angeles: The Center of an Agricultural Empire*. Los Angeles: Agriculture Department of the Los Angeles Chamber of Commerce, 1929.
24. McCawley, William. *The First Angelinos: The Gabrielino Indians of Los Angeles*. Banning, California: Malki Museum Press/ Ballena Press Cooperative Publication, 1996.
25. McWilliams, Carey. *Southern California: An Island on the Land*. First published in 1946 by Duell, Sloan & Pearce. Layton, Utah: Gibbs Smith, Publisher, 2010.
26. ____. *Factories in the Field: The Story of Migratory Farm Labor in California*. First published in 1939 by Little, Brown and Co.. Berkeley: University of California Press, 2000.
27. Miller, Howard A. *What the Newcomer Should Know About Agriculture in Los Angeles County and Southern California*. Los Angeles: Los Angeles Chamber of Commerce and Los Angeles County Board of Supervisors, 1940.
28. Monroy, Douglas. *Thrown among Strangers: The Making of Mexican Culture in Frontier California*. Berkeley: University of California Press, 1990.
29. Nadeau, Remi. *Los Angeles: From Mission to Modern City*. New York: Longmans, Green and Co., 1960.
30. Padilla, Victoria. *Southern California Gardens: An Illustrated History*. Berkeley: University of California Press, 1961.
31. Pauley, Kenneth E., and Carol M. Pauley. *San Fernando, Rey De Espana: An Illustrated History*. Spokane: Arthur H. Clark Co., 2005.
32. Pinney, Thomas. *A History of Wine in America: From the Beginnings to Prohibition*. Berkeley: University of California Press, 1989.

33. Reuther, Walter, H.J. Webber, and L.D. Batchelor. *The Citrus Industry, Volume I*. Berkeley: University of California, Division of Agricultural Sciences, 1967.
34. Ridgeway, Peggi, and Jan Works. *Sending Flowers to America: Stories of the Los Angeles Flower Market and the People Who Built an American Floral Industry*. Los Angeles: American Florists Exchange, Ltd., 2008.
35. Robinson, W. W. *Land in California: The Story of Mission Lands, Ranchos, Squatters, Mining Claims, Railroad Grants, Land Scrip, Homesteads*. Berkeley: University of California Press, 1948.
36. ———. *Ranchos Become Cities*. Pasadena: San Pasqual Press, 1939.
37. Roderick, Kevin. *The San Fernando Valley: America's Suburb*. Los Angeles: Los Angeles Times Books, 2001.
38. Sackman, Douglas C. *Orange Empire: California and the Fruits of Eden*. Berkeley: University of California Press, 2005.
39. Schroeder, C.A. "Recollections of Avocado History at U.C.L.A." *California Avocado Society 1992 Yearbook* 76 (1992).
40. Settle, Glen A., ed. *Here Roamed the Antelope*. Rosamond, California: Kern-Antelope Historical Society, 1963.
41. Starr, Kevin. *Americans and the California Dream 1850–1915*. New York: Oxford University Press, 1973.
42. Street, Richard Steven. *Beasts of the Field: A Narrative History of California Farmworkers, 1769–1913*. Stanford, California: Stanford University Press, 2004.

Chapter-Specific References:

Introduction

1. Cohen, Reut. "Obesity Rising in Los Angeles County." *1st and Spring* (blog), KCET, Oct. 26, 2011, http://www.kcet.org/news/1st_and_spring/public-health/obesity-rising-in-los-angeles-county.html .
2. Goodman, Jeff. "Hidden Hunger on the Rise." *Santa Monica Daily Press*, July 24, 2015.
3. Hamilton, Walter. "Income Disparity Is Wide in L.A." *LAT*, Feb. 22, 2014.
4. Holland, Gale. "L.A. County Tops List of Counties with Children at Risk of Hunger." *LAT*, June 14, 2013.
5. Holland, Gale, and Emily Alpert. "Poverty Kept Rising in Los Angeles after Downturn, New Data Show." *LAT*, Sept. 18, 2013.
6. Los Angeles County Department of Public Health. "Obesity and Related Mortality in Los Angeles County." In *A Cities and Communities Health Report*, 16. Los Angeles: Los Angeles County Department of Public Health, 2011.
7. Masters, Nathan. "The Sycamores of Southern California: A Brief History." *L.A. as Subject* (blog), KCET, June 20, 2013, http://www.kcet.org/updaily/socal_focus/history/la-as-subject/the-sycamores-of-southern-california-a-brief-history.html .
8. United States Census Bureau. "State and County QuickFacts." Accessed Oct. 30, 2015. http://quickfacts.census.gov/qfd/states/06/06037.html.

Chapter 1

1. Crespi, Juan. "The Naming of Santa Ana and Los Angeles." Vol. 1 of *The Literature of California: Writings from the Golden State*, edited by Jack Hicks, James D. Houston, Maxine Hong, and Al Young. Berkeley: University of California Press, 2000.
2. Driggs, Janet O., ed. *Not a Cornfield: History/Site/Document Essays + Timeline*. Los Angeles: Not a Cornfield, LLC, 2007.

3. Heizer, Robert F., ed. *The Indians of Los Angeles County: Hugo Reid's Letters of 1852*, Southwest Museum Papers, vol. 21. Los Angeles: Southwest Museum, 1968.
4. Sanchez, Jesus. "L.A.'s Cornfield Row: How Activists Prevailed." *LAT*, Apr. 17, 2001.

Chapter 2

1. Bancroft, Hubert H. *History of California Volume II*. San Francisco: A.L. Bancroft & Co. Publishers, 1885.
2. Barrows, H.D. "Early Governors of Alta California." *Annual Publication of the Historical Society of Southern California and Pioneer Register, Los Angeles* 4, no. 3 (1899).
3. Dunmire, William W. *Gardens of New Spain: How Mediterranean Plants and Foods Changed America*. Austin: University of Texas Press, 2004.
4. Mosk, Sanford A. "Subsidized Hemp Production in Spanish California." *Agricultural History* 13, no. 4 (Oct. 1939).
5. Nunis, Doyce B. Jr., ed. *The Founding Documents of Los Angeles: A Bilingual Edition*. Los Angeles: Historical Society of Southern California, 2004.
6. Rasmussen, Cecilia. "Shaman and Freedom-Fighter Led Indians' Mission Revolt." *LAT*, June 10, 2001..
7. Stoltze, Frank. "L.A. City Voters Could Dramatically Reduce Access to Medical Marijuana." *Represent!* (blog), KPCC Southern California Public Radio, May 20, 2013. http://www.scpr.org/blogs/politics/2013/05/10/13616/la-city-voters-could-dramatically-reduce-access-to/ .

Chapter 3

1. Barrows, H.D. "Don Antonio Maria Lugo; A Picturesque Character of California." *Annual Publication of the Historical Society of Southern California* 3 (January 1896).
2. Bell, Horace. *Reminiscences of a Ranger: Or, Early Times in Southern California*. Los Angeles: Yarnell, Caystile, & Mathes, Printers. 1881.
3. Dana, Richard Henry Jr. *Two Years before the Mast: A Personal Narrative of Life at Sea*. First published in 1840 by Harper and Brothers. Los Angeles: Ward Ritchie Press, 1964.
4. Daniel, Carrie R., Amanda J. Cross, Corinna Koebnick, and Rashmi Sinha. "Trends in Meat Consumption in the United States." *Public Health Nutrition* 14, no. 4 (Apr. 2011).
5. Parker-Pope, Tara. "Switching to Grass-Fed Beef." *Well* (blog), *New York Times*, Mar. 11, 2010, http://well.blogs.nytimes.com/2010/03/11/switching-to-grass-fed-beef/ .
6. Rasmussen, Cecilia. "Ex-Pirate Evolved Into a Leading Citizen." *LAT*, Feb. 13, 2000.
7. Wright, Doris Marion. *A Yankee in Mexican California: Abel Stearns, 1798–1848*. Santa Barbara: Wallace Hebberd, 1977.

Chapter 4

1. Engstrand, Iris W. *William Wolfskill, 1798–1866; Frontier Trapper to California Ranchero*. Western Frontiersmen Series, no. 12. Glendale, California: A.H. Clark Co., 1965.
2. Hutchison, Claude B. *California Agriculture*. Berkeley: University of California Press, 1946.
3. Lorenz, Anthony J. "Scurvy in the Gold Rush." *Journal of the History of Medicine and Allied Sciences* 12, no. 10 (1957).
4. Macconnell, Scott. "Jean-Louis Vignes: California's Forgotten Winemaker." *Gastronomica: The Journal of Critical Food Studies* 11, no. 1 (Spring 2011).

5. O'Neil, Stephanie. "Food Forward Volunteers Pick Backyard Fruit for the Needy." *Local* (blog), KPCC Southern California Public Radio, Dec. 22, 2011. http://www.scpr.org/news/2011/12/22/30485/food-forward-volunteers-pick-backyard-fruit-needy/ .
6. Ortlieb, Patricia, and Peter Economy. *Creating an Orange Utopia: Eliza Lovell Tibbets and the Birth of California's Citrus Industries*. West Chester, Pennsylvania: Swedenborg–Foundation Press, 2011.
7. Rasmussen, Cecilia. "Wine Industry Took Root on Olvera Street." *LAT*, May 28, 2000.
8. Wine Institute. "California Wine Profile 2012." Accessed Oct. 31, 2015. http://www.wineinstitute.org/files/CA%20Wine%20Stat%20Profile%202012.pdf .

Chapter 5

1. Curwen, Thomas. "A History of Paradise." *LAT*, Apr. 30, 2006.
2. Fickle, Tara. "A History of the Los Angeles City Market: 1930–1950." *Gum Saan Journal* 32, no. 1 (2010).
3. Nadeau, Remi A. *City-Makers, the Men Who Transformed Los Angeles from Village to Metropolis During the First Great Boom, 1868–76*. Garden City, New York: Doubleday, 1948.
4. Netz, Joseph. "The Great Los Angeles Real Estate Boom of 1887." *Annual Publication of the Historical Society of Southern California* 10, no. 1/2 (1915–1916).
5. Nordhoff, Charles. *California: For Health, Pleasure, and Residence. A Book for Travellers and Settlers*. New York: Harper & Brothers, 1872.
6. Padilla, Steve. "The Glory That Was Lang: 'Golden Spike' Linked Los Angeles, San Francisco Rail Lines at Station." *LAT*, Nov. 26, 1989.
7. Robinson, John W. *Los Angeles in Civil War Days, 1860–1865*. Norman: University of Oklahoma Press, 2013.
8. "The Last Rail: The Excursionists on Route to Los Angeles." *Sacramento Daily Union*, Sept. 5, 1876.
9. "The Land-Boom Runs Mad." *Chicago Tribune*, Feb. 8, 1888.

Chapter 6

1. Albertine, Susan. "Self Found in the Breaking: The Life Writings of Harriet Strong." *Biography* 17, no. 2 (Spring 1994).
2. Apostol, Jane. "Harriet Russell Strong: Horticulturalist, Conservationist, and Feminist." *California History* 85, no. 2 (2008).
3. California State Board of Horticulture. *Seventh Biennial Report of the State Board of Horticulture of the State of California*. Sacramento: State Board of Horticulture of the State of California, 1901.
4. Groff, Frances A. "Frank Wiggins." *The American Magazine*, Nov. 1910.
5. Ramos, David E., ed. *Walnut Production Manual*. Oakland: University of California Agriculture and Natural Resources, 1998.
6. Scheuring, Ann F. *Science and Service: A History of the Land-Grant University and Agriculture in California*. Oakland: University of California Division of Agriculture and Natural Resources, 1995.
7. Shaker, Jana. "The Founding and First One Hundred Years of the Citrus Experiment Station." *Chronicle of the University of California* 8 (Fall 2006).
8. "She's Boss of Ranch." *Chicago Inter-Ocean*, Oct. 26, 1896.
9. "Statistical Report of the California State Board of Agriculture." Sacramento: California State Board of Agriculture, 1913.

10. Strong, H.W.R. "Pampas Grass and Pomegranate in California." *Overland Monthly*, Apr. 1895.
11. U.S. Department of Agriculture Economic Research Service. "Food Access Research Atlas." Accessed Nov. 1, 2015. http://www.ers.usda.gov/data-products/food-access-research-atlas.aspx .
12. Willard, Charles Dwight. *A History of the Chamber of Commerce of Los Angeles, California: From Its Foundation, Sept., 1888, to the Year 1900*. Los Angeles: Kingsley-Barnes & Neuner, 1899.

Chapter 7

1. "Anaheim Ostrich Farm: Dr. Sketchley's Experiment in California." *New York Times*, Oct. 28, 1883.
2. Earnest, Francis V. "Saved from Death in Alligator's Jaws: Child's Uncle Forces Reptile to Let Go of Mangled Hand." *LAT*, July 10, 1925.
3. Emery, Christine. "Our Biggest Birds: A Glimpse of a Freak Crop—Ostriches." *LAT*, Feb. 23, 1930.
4. "Famed Gay's Lion Farm at El Monte to Close: Institution Has Been Visited by Million Persons." *LAT*, Dec. 25, 1942.
5. "Finger Is Added to an Alligator's Breakfast Food." *LAT*, Aug. 8, 1921.
6. Masquelier, Odile. "Paul De Longpre: An American Painter from Lyon." *Rosa Mundi* 24, no. 1 (Spring/Summer 2010).
7. Rinehart, John C. "Animal 'Farms' Attract Many: Fantastic Places Visited by Thousands Yearly." *LAT*, Feb. 23, 1930.
8. "Speaking of 'Unusual' Crops, the Southland Has 'Em." *LAT*, Feb. 12, 1928.
9. Stein, Sarah A. *Plumes: Ostrich Feathers, Jews, and a Lost World of Global Commerce*. New Haven, Connecticut: Yale University Press, 2008.
10. Warren, Viola Lockhart. "The Eucalyptus Crusade." *Southern California Quarterly* 44, no. 1 (Mar. 1962).

Chapter 8

1. "Bumper Crop Reports Speed Realty Sales." *LAT*, Oct. 13, 1929.
2. *Calavo, the Aristocrat of Salad Fruit*. Los Angeles: Calavo Growers, 1937.
3. Gast, Ross H. "Who'll Grow the Vegetables When the Jap Moves On?" *LAT*, Feb. 5, 1922.
4. Gerber, Judith. *Farming in Torrance and the South Bay*. Images of America. Charleston, South Carolina: Arcadia Publishing, 2008.
5. Hodgson, R.W. *The California Avocado Industry*. Berkeley: University of California College of Agriculture, 1930.
6. "Straight Outta Compton. . .On Horseback." *Around the Nation* (blog), National Public Radio, Apr. 3, 2011, http://www.npr.org/2011/04/03/134981907/straight-outta-compton-on-horseback .
7. Richardson, Russell. "Intensive Farming Brings City Living Conditions into the Country." *LAT*, Jan. 2, 1930.
8. Simmons, Ann. "Farming in Compton's Core." *LAT*, Feb. 20, 2011.
9. "$650 Buys an Acre Home Site with City Advantages at Richland Farms." Display Advertisement 209, *LAT*, Nov. 12, 1911.
10. "Under the Mighty Aqueduct a Fruitful Valley Unfolding." *LAT*, July 31, 1915.
11. Weeks, Charles. "The Charles Weeks Poultry Colony at Owensmouth, California." *The Pacific Poultryman*, Dec. 1922.
12. "What Publicity Did for Watts: Suburban Farm Section of Los Angeles, a Few Years Ago, Has Grown to Be a City of Three Thousand within Five Years." *LAT*, Sept. 11, 1910.
13. "Wilson Popenoe: In Appreciation." *California Avocado Society 1942 Yearbook*, 11–12: California Avocado Society 27 (1942).

Chapter 9

1. "Almost Free Berries Lure Hundreds to Fields in El Monte Area Strike." *LAT*, June 30, 1933.
2. "Are the Dec. 5 Fast-Food Strikes Happening in Your Town?" *Ms.cellany* (blog), *Ms. Magazine*, Dec. 4, 2013. http://msmagazine.com/blog/2013/12/04/are-the-dec-5-fast-food-strikes-happening-in-your-town/ .
3. Balderrama, Francisco E., and Raymond Rodriguez. *Decade of Betrayal: Mexican Repatriation in the 1930s*. Albuquerque: University of New Mexico Press, 2006.
4. "Berry Strike Gets Violent." *LAT*, June 10, 1933.
5. Bowling, Harry F. "El Monte, the Land of Free Water." *LAT*, Mar. 12, 1911.
6. Carriker, Robert M. *Urban Farming in the West: A New Deal Experiment in Subsistence Homesteads*. Tucson: University of Arizona Press, 2010.
7. "Farm Strike Parley Set." *LAT*, Mar. 25, 1936.
8. "First Homestead Families: New Federal Aid Sites Occupied at El Monte." *LAT*, Aug. 27, 1935.
9. "500 Raspberry Pickers Meet; Plan Fight." *Western Worker*, June 5, 1933.
10. "Fugitive Aliens Seized in Drive." *LAT*, Feb. 15, 1931.
11. Gast, Ross H. "A Review of the Year's Progress in Agriculture." *LAT*, Jan. 2, 1930.
12. Gazzar, Brenda, and Jason Henry. "San Gabriel Valley's Biggest Cities See Few Changes to Income, Poverty Rate in Last Two Years." *San Gabriel Valley Tribune*, Sept. 18, 2013.
13. Hsu, Tiffany. "Fast-Food Workers Rally Nationwide for Higher Wages." *LAT*, Dec. 5, 2013.
14. Kerr, Clark, and Paul S. Taylor. "Self-Help Cooperatives in California." In *Essays in Social Economics*, 191–225. Berkeley: University of California Press, 1935.
15. Khokha, Sasha. "Amid Fields of Plenty, a Farmworker's Wife Struggles to Feed Her Family." *The Salt* (blog), National Public Radio, Dec. 18, 2013. http://www.npr.org/blogs/thesalt/ 2013/12/17/252004699/amid-fields-of-plenty-farmworkers-struggle-to-feed- their-families .
16. "Labor Pool Organized: Farm Federation Head Cites Plans to Prevent Loss in Vegetable Fields." *LAT*, May 11, 1936.
17. Lopez, Ronald. "The El Monte Berry Strike of 1933." *Aztlán* 1, no. 1 (Spring 1970).
18. Maclay, Kathleen. "Low-Wage Fast-Food Jobs Leave Hefty Tax Bill, Report Says." *UC Berkeley News* (blog), UC Berkeley, Oct. 5, 2013. http://newscenter.berkeley.edu/2013/10/15/low-wage-fast-food-jobs-leave-hefty-tax-bill-report-says/ .
19. "Migrant Influx Laid to Farms: Short-Time Labor Demand Blamed for Transient Problem." *LAT*, Dec. 11, 1938.
20. Panunzio, Constantine M., Wade E. Church, and Louis Wasserman. *Self-Help Coöperatives in Los Angeles*. Berkeley: University of California Press, 1939.
21. Quin, Michael. "Hail! Fighting Berry Pickers!" *Western Worker*, July 3, 1933.
22. Rasmussen, Cecilia. "The Heyday and Decline of a Lively Barrio." *LAT*, Nov. 27, 1995.
23. "Raspberry and Potato Pickers Answer Union Call." *Western Worker*, June 12, 1933.
24. Starr, Kevin. *Endangered Dreams: The Great Depression in California. Americans and the California Dream.* New York: Oxford University Press, 1996.
25. Steinbeck, John. *The Harvest Gypsies: On the Road to the Grapes of Wrath*. Berkeley: Heyday Books, 1988.
26. "Striking Berry Picker Admits Beating Worker." *LAT*, June 23, 1933.
27. "Tear Gas Bombs Rout 300 Strike Rioters in Venice." *LAT*, Apr. 25, 1936.
28. "Threat to Burn Feed Hurled in Three-Day Dairy Strike." *LAT*, July 28, 1937.
29. Wollenberg, Charles. "Race and Class in Rural California: The El Monte Berry Strike of 1933." *California Historical Quarterly* 51, no. 2 (Summer 1972).

Chapter 10

1. Ainsworth, Ed. "Farm Output Rising Fast: Survey Shows County All-out in Response to 'Freedom Food' Drive." *LAT*, Apr. 26, 1942.
2. "Alhambrans Given Goal of 6000 Victory Gardens." *LAT*, Feb. 2, 1944.
3. Baumberger, Fred. "Valley Growers, Rancher in Last Ditch Battle against Urbanization." *LAT*, Oct. 4, 1959.
4. "Bulldozers Carve Route of Santa Ana Freeway." *LAT*, May 7, 1950.
5. English, Bob. "Veteran Walnut Rancher Bows out as Tracts, Missiles, Smog Bow In." *LAT*, Jan. 27, 1957.
6. "$4,000,000 Residential Project Launched in East Whittier." *LAT*, Jan. 14, 1951.
7. "Freeway Pictured Lane of Beauty." *LAT*, July 8, 1951.
8. "Girl Office Workers and Wartime Harvest." *LAT*, Aug. 18, 1943.
9. "Girl Students Pick Tomatoes: Beverly Hills and City College Workers Aid in Valley Harvest." *LAT*, Oct. 25, 1942.
10. Hebert, Ray. "Los Angeles County—Land of Contrasts." *LAT*, Feb. 22, 1959.
11. Hillinger, Charles. "Dairy Valley Thriving under Incorporation." *LAT*, Oct. 2, 1956.
12. Hulse, Jerry. "Farmers Fight Back at Encroachment of City." *LAT*, Nov. 28, 1957.
13. ———. "Los Angeles Agricultural Empire Fading before March of Industry." *LAT*, Nov. 27, 1957.
14. Hurley, Ray. *United States Census of Agriculture: 1954.* Washington, D.C.: U.S. Bureau of the Census, 1956.
15. ———. *U.S. Census of Agriculture: 1959.* Washington, D.C.: U.S. Bureau of the Census, 1962.
16. "Japanese Give up Crops in San Fernando Valley." *LAT*, Apr. 8, 1942.
17. Jones, John B. *The Songs That Fought the War: Popular Music and the Home Front, 1939–1945.* Lebanon, New Hampshire: University Press of New England, 2006.
18. Murata, Alice. "Interview with Hiroshi Mayeda." *Regenerations Oral History Project: Rebuilding Japanese American Families, Communities, and Civil Rights in the Resettlement Era: Chicago Region: Volume 1.* Los Angeles: Japanese American National Museum, 1997.
19. Naden, Kenneth D. "Is Urbanization a Threat to Food Supply?" *LAT*, Dec. 28, 1954.
20. "Network of Japanese Farms Covers Vital Southland Defense Areas." *LAT*, Mar. 4, 1942.
21. "New Victory Aid Planned: F.S.A. Maps All-out Food Production for Nation." *LAT*, Jan. 19, 1942.
22. "Program Involving 940 New Homes Announced." *LAT*, Oct. 1, 1950.
23. "Realty Board Aids Garden Movement." *LAT*, Mar. 21, 1943.
24. "Southland Launches Victory Garden Drive." *LAT*, Mar. 1, 1945.
25. "Tomato Crop to Be Saved: Civic Leaders, Officials and Women Will Join with Students in Harvest." *LAT*, Sept. 24, 1942.
26. "Victory Garden Plan Intensified." *LAT*, Jan. 25, 1943.
27. "Victory Garden Seed Planting Drive Pushed." *LAT*, Apr. 12, 1943
28. Whyte, William H., ed. *The Exploding Metropolis.* 3rd ed. Classics in Urban History. Berkeley: University of California Press, 1993.

Chapter 11

1. "Back-to-the-Soil Fever Hits Suburbanites." *LAT*, Jan. 25, 1976.
2. Balter, Michael. "Drops of Mistrust Cloud Malathion Spray." *LAT*, Aug. 11, 1989.
3. Booth, William. "DDT Dump Off California Coast Poses Unique Cleanup Challenge." *San Francisco Chronicle*, Sept. 20, 2011.

4. Brown, Allison. "Farmers' Market Research 1940–2000: An Inventory and Review." *American Journal of Alternative Agriculture* 17, no. 4 (2002).
5. Burleigh, Irv. "Lorenzen Mounts 'Victory Garden' War on Bradley." *LAT*, Aug. 7, 1975.
6. Carey, James. "The Mediterranean Fruit Fly in California: Taking Stock." *California Agriculture* 46, no. 1 (Jan.–Feb. 1992).
7. Cohen, Jerry. "Agricultural Land Is Growing Scarce: Nursery Plants Are Now the Top Crop in Los Angeles County." *LAT*, May 10, 1987.
8. Dunn, Ashley. "Aerial Use of Malathion Is Questioned: Pest Experts' Report Says State Should Reconsider Spraying of Pesticide over Neighborhoods Because of Potential Health Hazards." *LAT*, Mar. 27, 1991.
9. Fairbanks, Robert. "Brown Angers Agricultural Leaders: Medfly Aerial Spraying Barred." *LAT*, July 9, 1981.
10. Feldman, Paul. "Cast of 'Cheers' Leads Westwood Protest against Medfly Spraying." *LAT*, Jan. 24, 1990.
11. Fleming, Louis. "Air Pollution Menaces Richest Farm Areas." *LAT*, Jan. 8, 1961.
12. Holguin, Rick. "Mooove 'Em Out: Southeast L.A. County Once Had Hundreds of Dairies, but Today 3 Survive." *LAT*, Mar. 5, 1992.
13. La Ganga, Maria L. "Farmers Try to Polish up Their Tarnished Image." *LAT*, Nov. 4, 1990.
14. Malnic, Eric. "Spray Zone Grows with New Medfly Find in L.A. Area." *LAT*, Oct. 13, 1981.
15. Nakata, J.K., H.G. Wilshire, and G.G. Barnes. "Origin of Mojave Desert Dust Plumes Photographed from Space." *Geology* 4 (Nov. 1976).
16. O'Reilly, Richard. "Brown Bows to Ultimatum: Orders Aerial Spraying, Blasts Quarantine Threat." *LAT*, July 10, 1981.
17. O'Shaugnessy, Lynn. "Last Chance Peach Stirs Doubt in Antelope Valley: Investors Say Fruit Is Tasty but Overrated." *LAT*, Apr. 13, 1987.
18. Parsons, Russ. "The Idea That Shook the World." *LAT*, May 24, 2006.
19. Schneider, Iris. "Bountiful Harvest-City Style: Both Buyers and Sellers Love Going to Farmers' Markets." *LAT*, Dec. 15, 1985.
20. Stevenson, Richard. "7 Copters Face Off in Medfly Dispute." *New York Times*, Feb. 23, 1990.
21. Tracy, Phil. "Silent Spring in Echo Park." *LAT*, Apr. 8, 1984.
22. "UCLA May Move Farm Activities." *LAT*, Oct. 12, 1958.
23. Wallace, Charles P. "Last Rites for the Medfly: Champagne, Handshakes." *LAT*, Sept. 22, 1982.

Epilogue

1. Agha, Marisa. "Urban Beekeeping Raises Southern California Concerns." *Sacramento Bee*, Feb. 14, 2015.
2. Arnold, Shayna Rose. "The Urban Garden Movement Keeps Growing: City Council OKs Edible Plants for L.A. Parkways." *Los Angeles Magazine*, Mar. 6, 2015.
3. Becerra, Hector. "L.A. Garden Shut Down: 40 Arrested." *LAT*, June 14, 2006.
4. Benjamin, Alison. "Urban Bees Fare Better Due to Varied Diet, Research Reveals." *Guardian*, Aug. 17, 2010.
5. Boxall, Bettina. "Bills Regulating State's Groundwater Not an Instant Fix for Aquifers." *LAT*, Sept. 7, 2014.
6. Butterfield, Bruce. *Garden to Table: A 5-Year Look at Gardening in America.* Williston, Vermont: National Gardening Assoc., 2014.
7. Farr, Jory. "Judge Halts Eviction of Urban Gardeners-South Central Los Angeles Group Wins Injunction against City." *Los Angeles Daily Journal*, Mar. 29, 2004.
8. Fernald, Marcia. *America's Rental Housing: Evolving Markets and Needs.* Cambridge, Massachusetts: Joint Center for Housing Studies of Harvard University, 2013.

9. Finley, Ron. TED. "A Guerilla Gardener in South Central LA." Filmed Feb. 2013. TED video, 10:45. https://www.ted.com/talks/ron_finley_a_guerilla_gardener_in_south_central_la .
10. Fox, Sue. "Farm Growth Rooted in Carrots; Agriculture: The Orange Vegetable Has Spearheaded an Increase of Cropland in L.A. County's High Desert." *LAT*, Aug. 26, 2001.
11. Gambino, Megan. "Turning Shipping Containers into Urban Farms." *Smithsonian.com*, Dec. 9, 2014, http://www.smithsonianmag.com/innovation/turning-shipping-containers-into-urban-farms-180953295 .
12. Garrison, Jessica. "Roosters and Chickens and Pigs in Urban L.A., Oh My!" *LAT*, May 25, 2008.
13. Harvard School of Public Health Nutrition Source. "What Should I Eat: Vegetables and Fruits." Accessed Nov. 8, 2015. http://www.hsph.harvard.edu/nutritionsource/what-should-you-eat/vegetables-and-fruits/ .
14. Jackson, Jaemi, Kelly Rytel, India Brookover, Nicholas Efron, Genevieve Hernandez, and Erik Johnson. *Cultivate L.A.: An Assessment of Urban Agriculture in Los Angeles County*. Los Angeles: University of California Los Angeles, Dept. of Urban Planning, 2013.
15. Jennings, Angel. "L.A. City Council Introduces Plan to Encourage Urban Farming." *LAT*, Oct. 8, 2014.
16. Kennedy, Scott Hamilton. *The Garden*. Black Valley Films, 2008. Film, 80 minutes.
17. Lee, Linda. "Adjudication Annoys Both Sides: Water Fight Lasts for a Decade." *Antelope Valley Press*, July 22, 2008.
18. Lehner, Peter. "Planting Food and Hope: The Inspiring Afterlife of South Central Farm." *Switchboard* (blog), Natural Resources Defense Council, May 2, 2013, http://switchboard.nrdc.org/blogs/plehner/planting_food_and_hope_the_ins.html.
19. Lopez, Steve. "In the Weeds of Bureaucratic Insanity There Sprouts a Small Reprieve." *LAT*, Aug. 20, 2011.
20. Low, Sarah, Aaron Adalja, Elizabeth Beaulieu, Nigel Key, Steve Martinez, Alex Melton, and Agnes Perez. *Trends in U.S. Local and Regional Food Systems: Report to Congress*. Washington, D.C.: U.S. Department of Agriculture Economic Research Service, 2015.
21. Morier, Douglas. *Social Determinants of Health: Rising Food Insecurity in Los Angeles County*. Los Angeles: County of Los Angeles Public Health, 2015.
22. Morin, Monte. "Medfly Entrenched in California, Study Finds." *LAT*, Aug. 6, 2013.
23. Nagourney, Adam. "California Imposes First Mandatory Water Restrictions to Deal with Drought." *New York Times*, Apr. 1, 2015.
24. Peterson, Lars. "Can You Really Save Money on Eggs by Raising Chickens at Home?" *My Money* (blog), U.S. News and World Report, Mar. 25, 2015. http://money.usnews.com/money/blogs/my-money/2015/03/25/can-you-really-save-money-on-eggs-by-raising-chickens-at-home .
25. Rogers, Paul. "California Drought: After Years of Overpumping Groundwater, State May Be Ready for Reforms." *San Jose Mercury News*, Mar. 20, 2014.
26. Seidenwurm, Amy. "Pecks in the City." *Los Angeles Times Magazine*, Mar. 29, 2010.
27. Shevory, Kristina. "The Beekeeper Next Door." *New York Times*, Dec. 8, 2010.
28. "Top 100 Cultural Trends Shaping the West." *Outdoor Adventure* (blog), Sunset.com, Feb. 1, 2011, http://www.sunset.com/travel/outdoor-adventure/cultural-trends-sunset-magazine-0211/cultural-trends-chicken-coop-0211 .
29. Wagner, Laura. "Backyard Beekeeping Approved in Los Angeles." *The Two-Way: Breaking News from NPR (blog),* National Public Radio, Oct. 14, 2015, http://www.npr.org/sections/thetwo-way/2015/10/14/448725988/backyard-beekeeping-approved-in-los-angeles .
30. Xia, Rosanna, and Diana Marcum. "California Braces for a Deadly Stalker of Citrus." *LAT*, Apr. 14, 2012.
31. Zahniser, David. "L.A. Council Clamps Down on a Crowing Problem: Roosters." *LAT*, Sept. 22, 2009.

Acknowledgments

We want to thank Angel City Press publishers Paddy Calistro and Scott McAuley, who believed in our vision for this book and guided us as we worked with them on this project over several years. We are deeply grateful for their support. Thanks also to graphic designer Amy Inouye of Future Studio for making the book beautiful, and to editor-extraordinaire Niree Perian for helping us to bring the book across the finish line.

This book would not be as photographically diverse or compelling without the assistance from the many places and people who so generously shared their historic images and ephemera with us. We would like to give a special thank-you to the following individuals for going above and beyond the call of duty to help us: John Cahoon, Seaver Center for Western History Research, Los Angeles County Museum of Natural History; Glen Creason, Los Angeles Public Library; Javier Garibay, Special Collections University of California, Los Angeles; Terri Garst, Los Angeles Public Library; Gloria Hillard, Contributing Correspondent, NPR; Donald Hodel, U.C. Cooperative Extension; Laura Jellum, Food Forward; Brent Riggs, Seaver Center for Western History Research, Los Angeles County Museum of Natural History; Bree Russell, Special Collections, University of Southern California; Sherry Smith, Bolton Hall Museum; Paul R. Spitzzeri, Workman and Temple Family Homestead Museum; Dace Taube, Regional History Librarian, University of Southern California; and Cindy Werner, Los Angeles County Agricultural Commissioner/Weights and Measures.

—R.S. and J.G.

I would like to thank my employers at U.C. Cooperative Extension, part of the University of California's Division of Agriculture and Natural Resources. Their support was instrumental, especially that of my supervisor, Keith Nathaniel, whose encouragement is deeply appreciated. I also want to thank my family who not only tolerated, but encouraged the weekends, evenings, and conversations dominated by "the book," including my sons Jack and Roy Mabie, my father Frederic Surls, and my sister Rebecca Mooty.

—R.S.

I also thank my family who spent years hearing about "the book," and offered moral support and encouragement when I needed it the most. My deep gratitude to my husband Dan Hylands, my father Joseph Gerber, my mother Joanne Feese, and my sisters Sue Matsui and Miriam Mann.

—J.G

Image Credits

The authors and Angel City Press are grateful to the individuals and institutions that provided the images so critical to the history shared in *From Cows to Concrete*. Except as noted below, photographs, maps, and ephemera on these pages are from the extensive and comprehensive Photograph Collection and Map Collection of the Los Angeles Public Library. Without the Library's vast resources, the story of the rise and fall of farming in Los Angeles County would have been incomplete.

Karen Ard: 56.
Autry National Center, Los Angeles: Southwest Museum of the American Indian Collection, Autry Museum, 17.
Bowers Museum, Santa Ana, California: 16.
California Historical Society: 72; California Historical Society Collections at the Autry National Center, 2–3.
California State University, Dominguez Hills, University Library Department of Archives and Special Collections: 8.
California Wine Institute: 26.
Cal Poly Pomona, University Library, Special Collections: 57.
Judith Gerber: 21, 85 (inset), 98, 100, 150 (left and center), 153.
Jacob Gutierrez, Keepers of Indigenous Ways: 14.
Gloria Hillard: 112.
Donald Hodel, University of California Cooperative Extension: 13.
The Huntington Library, Art Collections, and Botanical Gardens: 74, 95, 106.
Laguna Art Museum Collection; Gift of Nancy Dustin Wall Moure: 4–5.
Library of Congress: 6, 128, 140.
Los Angeles County Agricultural Commissioner/Weights and Measures: 162.
Midori Books: 125.
Palos Verdes Library District Local History Collection: 126.
Seaver Center for Western History Research, Los Angeles County Museum of Natural History: 40, 43, 62, 84, 85, 90, 93.
South Pasadena Public Library: 99.
Rachel Surls: 104, 138.
The Theodore Payne Foundation for Wild Flowers and Native Plants: 105.
University of California, Berkeley, Bancroft Library: 39, 48, 50, 94 (top), 139.
University of California, Los Angeles, *Los Angeles Daily News* Collection: 146 (top), 148.
University of California, Los Angeles, *Los Angeles Times* Collection: 144, 149, 158–159, 166 (bottom), 169, 170, 173.
University of California, Los Angeles, Special Collections: 58–59, 176.
University of Southern California, Doheny Memorial Library, Special Collections, California Historical Society Collection, 1860–1960: 29, 33, 36, 60–61, 64, 66, 82, 89, 92, 96, 108, 127, 130, 133, 154.
University of Southern California, Doheny Memorial Library, Special Collections, *Los Angeles Examiner* Collection: 123, 155.
Whittier Historical Society and Museum: 76.
Workman and Temple Family Homestead Museum: cover, 1.
Stephen Zeigler: 178.

Index

From Cows to Concrete: The Rise and Fall of Farming in Los Angeles
by Rachel Surls and Judith Gerber
ISBN 978-1-62640-031-3

Design by Amy Inouye, Future Studio

Library of Congress Cataloging-in-Publication Data is on file.

Published by Angel City Press
at Los Angeles Public Library
www.angelcitypress.com

Printed in Canada